The reflexive universe

also by Arthur M. Young

The Geometry of Meaning

The Bell Notes

Which Way Out? and other essays

Arthur M. Young

The reflexive universe

Evolution of consciousness

Robert Briggs Associates

Grateful acknowledgment is made for permission to use the following copyrighted material.

Excerpt from pp. 175–176 of *Clairvoyance and Materialization* by Gustav Geley: translated by Stanley Barth (1927). Used by permission of Ernest Benn Limited.

Chart of the plant kingdom from Harold C. Bold, *The Plant Kingdom*: 3rd edition, © 1970. Reprinted by permission of Prentice-Hall, Inc., Englewood Cliffs, N.J.

Illustrations on pp. 100, 101, 102, 121, and 103 from Villee, C. A.: *Biology*, 6th edition: Philadelphia, W. B. Saunders Company, 1972. Reproduced by permission of the author and publisher.

Illustration on p. 113 of sunflower with counterclockwise and clockwise spirals: Adapted from "Mathematical Games" by Martin Gardner. Copyright © 1969 by Scientific American, Inc. All rights reserved.

Manufactured in the United States of America
Designed by Jerry Tillett
Sixth Printing—1990

Robert Briggs Associates
400 Second Street #108
Lake Oswego, OR 97034

Library of Congress Cataloging in Publication Data

Young, Arthur M   1905–
The reflexive universe.

"A Merloyd Lawrence book."
Includes bibliographical references and index.
1. Evolution. 2. Science—Philosophy. 3. Consciousness. I. Title.
QH371.Y68   575.01   76–2675

ISBN 0–9609850–6–9 (paperback)

The Reflexive Universe
was originally published by
Merloyd Lawrence

The publishers wish to thank Sheila La Farge
for extensive editorial assistance
in preparing this book for publication.

Acknowledgments

In the course of rewriting this book a number of times, I have drawn on many friends for ideas, criticism, and suggestions—Ted Bastin, Chris Bird, Oscar Brunler, Jean duCharm, Chris Clark, John Coggeshall, Ira Einhorn, George Hall, Charles Hapgood, David Hill, Bartram Kelley, Sheila La Farge, Payson Loomis, Mary Benzenberg Mayer, Alice Morris, Charles Musès, Ken Pelletier, Charles Price, Jeanne Rindge, Ivan Sanderson, Joan Schleicher, Eric Schroeder, Alice Schwarr, Saul Paul Sirag, Harry Smith, Vincent Smith, Nick Spies, Kelvin Van Nuys, Chycherle Waterston, Rick Werner, Chris Young, and most of all my wife, Ruth—each has made a distinct contribution.

Contents

Chapter VIII | The plant kingdom   *90*

Chapter IX | The animal kingdom   *110*

Chapter XVI | The evolution of the self   242

Appendix I | A brief outline of the theory   255

Appendix II | Seven-ness   259

Appendix III | The phase dimension (from Eddington's
***Fundamental Theory*)**   283

Introduction

It has become customary to express one's appreciation of a bold and original thinker by saying that his work defies classification. Arthur M. Young's writings are, however, quite readily classified and are no less original for all that. He belongs to that small class of modern men and women who attempt to think for themselves and to look at nature with their own eyes. This enterprise is part of a great and ancient tradition of the human race.

My intention here is neither to extol nor to criticize Arthur Young's ideas as such. What I wish to speak of is rather this whole question of looking with one's own eyes, an ideal that not only lies at the root of modern science but forms the foundation of all real philosophy, religion, and art through the ages. It is a very difficult enterprise and in our culture is far more esteemed in word than in deed.

Young is a scientist and a metaphysician. Like an increasing number of scientifically trained men and women in our society, he seeks to correct the imbalance that science has introduced into our lives through its denial of approaches to knowledge that rely on sources other than the evidence of the senses. Through this denial, science has reduced itself to scientism, and in reaction to this there has now appeared in our culture a strong current of interest in religious, mystical, and esoteric ideas about man and nature. As his book *The Bell Notes* indicates, Young entered this current at a time when it was hardly more than a rivulet. In these notes, which date from 1945 to 1947, when he was struggling to perfect his invention of the Bell helicopter, Young writes:

> What is our scientific civilization? Nothing? No, it is a very impressive something, but this something is really not what we thought it was . . .

Once upon a time there was magic. Civilization learned to reproduce, reduce to a formula, and duplicate a part of this magic. That is our world of science. The rest of the magic has been tossed on the junk heap. We are now scratching around trying to piece it together again.

Young writes of the world that modern science has revealed with much the same attitude that one finds in the great naturalists of the eighteenth and nineteenth centuries. He marvels as well as explains, a gift that the modern scientist is losing in the thicket of government funding and computer printouts, and the avalanche of mediocre, tedious, and sometimes even fraudulent research. Like the great naturalists, Young also marvels at his own feelings toward nature and seems to understand that these special feelings associated with the search for meaning are perhaps the single most important cognitive instrument of the true scientist.

His books invite us into a sacred universe. It is here that his work deserves to be measured against the achievements of the great metaphysicians of the premodern era. Strange as it may sound to the contemporary positivist philosopher, the fact is that the major metaphysical teachings of the ancient world were all founded on experience. Not the outer, external experience of the modern scientist, but the equally rigorous internal experience of the authentic contemplative.

There once existed a way of knowing that can be called *inner empiricism*. Through awareness of the movements and energies within the human microcosm, the great metaphysicians of the past were able to formulate ideas about the macrocosm—nature in all its universal vastness—that answered the need of the human mind not only for knowledge, but for meaning. When the contemplative, experiential aspect of metaphysics was lost and forgotten in the modern era, metaphysics degenerated into mere speculation, without roots in the world of hard inner facts. As such, it was deservedly castigated by the principal modern philosophers, such as Hume, Kant, and Wittgenstein.

In Arthur Young's work, however, we see a pioneering modern attempt to reclaim the world of inner facts as a world of equal reality to the world of outer facts. His work resoundingly echoes with the rediscovery of ancient truths—alchemical symbols, spiritual mathematics, the great hierarchical chain of consciousness and purpose—

while remaining faithful to the latest theories and discoveries of modern science. He respects the inner feeling for meaning quite as much as any solid scientific thinker respects the need for logic and precise external observation. And he shows us that the concepts of modern science simply cannot answer the questions of the inner world.

In short, Arthur Young is a man in search of ideas that satisfy both the subtle, inner perceptions of man's knowing heart and the outer demands of the functional, pragmatic mind. To find such ideas is perhaps the greatest need of our era—the need for a truly comprehensive worldview and not merely a loose set of hypotheses that arrange the data of the outer world. The outer world is only half of reality. If we are to cease being half human, we are going to have to begin by finding ideas that address both halves of reality itself. For this we need seekers like Arthur M. Young.

We need a new tradition of inner empiricism.

Jacob Needleman
San Francisco   1984

Preface

The purpose of this book is to develop a theory of the evolution of the "universe," and by universe we are referring principally to the one of which man is a part. This requires a foundation in which the laws of physics are integrated with the testimony consistently expressed by empirical evidence not explained by science, such as ESP, as well as by spiritual insights. It establishes a formal system in which consciousness, especially higher consciousness, can exist.

Science as usually interpreted does not provide for consciousness. Accordingly, there has arisen a conflict between science and religious thought, a conflict which is frequently dismissed because, it is said, science is not concerned with final issues. Such an interpretation of science has invaded other areas. This is particularly true of the social and human sciences, such as psychology. Here it is difficult to understand how we can have a valid science of man, if that science does not recognize final causes, since for man final causes are of the greatest significance. They are, in fact, the touchstone of his behavior. They inspire his works, his goals, his responsibilities to his fellows.

We have a situation in which science, modestly disclaiming any knowledge of the ultimate issues and sticking to its requirement of particles operated on by forces, has become the model. This model has in turn become the ideal of the psychologists who, dismissing all subtlety of the psyche as "metaphysical," create their image of man as a biological machine which can have no mind that is not brain, and no psyche that is not explicable as chemical activity, simply because, in their view, the laws of physics do not recognize any other fundamental

xix

ingredients on which to draw. This being so, we can no longer forgo an invasion of the inner sanctum of science to see if it is indeed true that the universe contains nothing but billiard balls (particles, drives, response mechanisms, etc.).

Origins of the theory

Having given the reader an idea of what the book is about, I would like to offer a brief description of the steps by which my theory came into being. This will also conveniently involve acknowledging my sources, which include cosmologies embodied in old myths. Despite current neglect, these ancient concepts are in close accord with modern scientific findings.

The beginning was my ambition, while still in college, to construct a theory of the universe. I was intrigued by the Einstein theory of relativity, and my request for a course in relativity was granted. I was also influenced by Bertrand Russell's speculations on logic in his *Introduction to Mathematical Philosophy.* *

The spirit of the times, emerging from the horse-and-carriage era, was rejoicing in "new thought." The automobile, the telephone, the radio, the motion picture inspired an enthusiasm for encompassing space. Lindbergh's solo flight to Paris expressed an emerging ambition to contain the universe mentally. The theory of relativity claimed to capture totality in the pattern of space and time.

My first attempt, in 1927, to construct a theory of the universe was influenced and motivated by the same confidence in the possibility of explaining the universe in intellectual terms. I called it the theory of structure; it was based on a theme similar to that of relativity, that reality could be formulated and captured in pattern. But almost as soon as I started it, I encountered a difficulty: the enigma of time, an issue I still feel is vital. I considered that the neglect of time was responsible for the Cretan paradox, discussed by Bertrand Russell: someone says that

* Russell, Bertrand. *Introduction to Mathematical Philosophy.* London: Allen & Unwin, 1919, 1960; New York: Macmillan Co., 1919.

all Cretans are liars; but he himself is a Cretan. If he speaks the truth, then he is lying; if he lies, he is speaking the truth. The problem, I felt, could be resolved only by making logic subordinate to time, and this required expanding my theory to include time. A distinction must be made between an act in process and an act completed—such as a verb for the former, and a noun for the latter—which could include verbs in the past tense. The Cretan could say what he wished about past statements, but he could not do so for the statement he was making, for the very practical reason that it was not yet complete: judgments could not include themselves.

I concluded that structure could not properly portray time. Structure is a system of relationships that exist all at once, in simultaneity. It cannot encompass that which takes time to unfold. I changed my theory to one of *time–structure*, and then to the *theory of process*. This was to be more general than the theory of structure: it would include structure as a moving picture includes the individual frame.

Relativity had, of course, included time as a fourth dimension, and though it gave time an imaginary coefficient in the formula for interval, this did not impress me as a sufficient distinction. Relativity still referred to the "structure" of space–time, which I felt overlooked the crux of that which distinguishes time from space: its asymmetry and one-wayness.

I had noticed in art that the insistence on symmetry came at a point

Tang (8th-century) bronze   Later bronze (10th to 13th century)

in time when the particular art form was beginning to deteriorate. In the development of Chinese vases, for example, the curves of the earlier forms are asymmetrical and later give place to forms in which the bounding line is the sine curve. The outlines of such vases coincide if one is inverted, indicating symmetry. While this is "pleasing to the eye" as a curve, it becomes vacuous as a statement. The same curve, known as the Hogarth line of beauty, came to the West also at a time when formal art was becoming exhausted.

So, the appeal of symmetry, of questionable validity in art, could be misleading for a theory of the universe and erroneous in the treatment of time because it disregards "time's arrow" (one-wayness). I intended that my theory of process should stress that very quality of time that the theory of relativity ignored.

I later learned that, soon after this, in 1928, Dirac had discovered a mathematical expression that predicted the electron and the as yet undiscovered positron (which, like the heavenly twin in the Egyptian story, remains on the other side of the river of rebirth). Dirac's formula, unlike the quadratic expressions of relativity, was linear. A linear equation employs plus and minus signs and thus preserves the asymmetry that a quadratic expression, which is a sum of squares, factors out (because the square of a minus is a plus). The asymmetry registered by Dirac reveals the hidden positron.

But to return to my story. I found I could carry my theory of process no further. (It had not yet occurred to me that process had certain definite stages, nor that each of these stages had a distinct character.) At that point my life took a new direction; I decided that I must learn to solve problems to which the answers could be tested, and so turned to invention. But what to invent? After some preliminary attempts based on a misunderstanding of aerodynamics, I became more circumspect and in late 1928 went to the patent files in Washington to evaluate some of the possibilities I'd been thinking about. Some of these, like sound on wire (now on tape), I found had already been invented. Others were not promising for other reasons. Choice fell on the helicopter, which at that time had a long history of failures and clearly needed doing.

I started on the helicopter, with which I was involved for nineteen years, the first twelve on my own, the last seven under the auspices of Bell Aircraft.

While I have always been intrigued by science, I do not consider myself a scientist. The scientist has a certain *attitude* toward nature. He is preoccupied with the discovery of law—and having discovered it, he holds it sacred. The inventor too must discover law, but this is not his goal. He has his mind set on something he wants to achieve, to fly, for example, or to communicate without wires. So he must both learn the law and then apply it, which involves a turnabout, a change of direction. The law is essentially restrictive, it limits the possible; but when it is stated objectively, we may find that it can be turned about and will, through its very certainty, provide the means by which our end can be achieved.

Here again is an example of time's arrow; if B is always preceded by A, we can say A causes B, and not otherwise: cause and effect depend on the direction of time. On the other hand, if we know that A causes B, and we observe B, we can then deduce A; "where there's smoke, there's fire." These two principles—causality and inference—make it possible to make determinism work for us, to make the laws of nature expand our freedom rather than restrain it.

This attitude of learning to *use* a law instead of being blocked by it may have played a part in my conception of determinism or law as the *agency* of free will rather than inherently in conflict with it. This is important for the theory of process, since process, as we will later show, must create determinism in order to have means to achieve its goal (see end of Chapter I).

Many problems in the development of the helicopter involved this sequence: finding what happens, the laws or regularities, and rerouting one's course to make the same laws work to help; as when the carpenter planing a board discovers he is working against the grain and turns the board around.

I also found in the helicopter an example of how evolution works. I found by experience that without purpose, without a goal-directed activity, the helicopter could not possibly have evolved.

Because of its analogous function of building from blueprints, we might liken the assembly line to DNA. Evolution, it is supposed, is due to accidental mutations in the DNA. But the helicopter assembly line brought home to me quite forcibly that there was a built-in predisposition in what formerly had been an airplane assembly line to resist the change to helicopters and revert to airplane manufacture.

This could be counteracted only by constant vigilance. As a crowning touch, when the manufacture was transferred to Texas and new equipment was required, the equipment ordered was that for the manufacture of airplanes. The error was caught in time, but by a person outside the manufacturing organization, one of the helicopter team.

Purpose is the important factor in developing a machine. The tendency of philosophers who know nothing of machinery to talk of man as a mere mechanism—intending by this to imply he is without purpose—shows a lack of understanding of machines as well as of man. Indeed, there never was a machine that did not have a purpose. And there is perhaps no purpose that does not require a machine, whether a human body or some other kind, to achieve it.

From the fact that examination of the physical parts of a machine, often referred to as the hardware, will not disclose the purpose unless assembled and in operation, it should be easy to infer that one cannot understand man by an examination of his physical organism, his body, alone.

When, in 1948, the Bell helicopter reached production status, I was free to return to my original interests. My release found me intrigued by such phenomena as precognitive dreams and the ability of the African violet to regenerate the whole plant from a piece of a leaf. These, and problems of ESP with which I then became absorbed, called for a more inclusive science than the physics of my college days.

I again took up my theory of process, not only for its emphasis on time, but for the presence implicit in process of a purposiveness that pushes toward the attainment of a goal.

I had also become interested in ancient myths, in particular the myths of cosmogony, which describe how the universe came into existence. In many of these accounts, process is described as occurring in *seven* stages. This is true of Genesis, in which God makes the universe in six days and rests on the seventh. As the Genesis account itself puts it, these were not days, but *generations.* So too the Zoroastrian and the Japanese accounts. Not all traditions explicitly state that there are seven stages, but by noting correspondences we can find the implication of seven stages in yet other myths: for example, the Greek myth of Cronus. Most emphatic in its insistence on seven-ness is the ancient Hindu tradition. In any case, I now had a hint, perhaps a directive, that *process involves seven stages.*

But as a scientist, of course, I could not blindly accept this number merely on the basis of tradition, for, like most modern minds, I suspected that mere superstition had led to its selection, rather than any valid theoretical reason which required seven stages and not some other number.

It was at this time that a friend, Harry Smith, reminded me of the well-known fact that the torus, or doughnut, has a unique topology, such that a map drawn on its surface requires seven colors in order for all bordering countries to be distinguished by differences in color. Now on the ordinary surface, that of a plane or a sphere, such a map requires no more than four colors. Since colors are distinctions, it occurred to me that, just as the sphere may be thought of as analogous to structure, so the torus may be *analogous to process*. If so, its seven colors would correspond to the seven stages which might be expected in *process*.*

Now the torus shape, which is also that of a vortex, occurs widely in natural phenomena: it is the shape of a magnetic field, of a tornado, and of eddies in water. And especially interesting is the fact that it is the only manner by which self-sustained motion can exist in a given medium. It is a unique manifestation of air in air (a tornado) or of water in water (an eddy or whirlpool). Of course, we can have waves on the surface of water, but that is at the boundary between two

The torus   Two circularities in the torus

---

* Even more impressive are the mathematical credentials of the torus. Mathematics recognizes topology, the science of surfaces, as dealing with more profound relationships than does geometry. Geometry deals with measure and angles of figures on a surface, but topology deals with different classes of surface, such as the Moebius strip or the torus (see Appendix II). This should be important because the abstractions of mathematics have often proved to apply to real situations; the imaginary numbers are invaluable in equations which deal with waves or oscillation; the curved space of Riemann gave Einstein the basis for relativity. Toroidal space–time could be next.

elements. The vortex is consubstantial with its matrix; they are the same material.

What first appealed to me about the torus was that I now had, from theoretical considerations of the most fundamental sort (topology or the science of surfaces), a possible sanction for *assuming* that process has seven stages. Such sanction is stronger than that from any empirical test, because individual instances can never guarantee an invariable rule; there is always the possibility of a white crow.

In this Introduction I will not stop to give further arguments connecting process with the topology of the torus. They are set forth in Appendix II. Here I am simply giving an account of the steps by which my theory developed. Suffice it to stress that I considered I now had reason to take seriously the ancient accounts of creation in seven stages and to accept this number as a working hypothesis.

This takes me to another influence, the *Mahatma Letters*,* a book put together by Sinnett from letters he claimed were received from one of the Masters who inspired the theosophical tradition. Although well written, it is not an easy book; I read it several times before I appreciated it. But it was quite explicit on the subject of the seven stages of evolution. The author spoke of the known animal, vegetable, and mineral kingdoms, to which he added a kingdom beyond animal and *three more preceding the mineral*. These, he said, he could not go into because science did not yet know of their existence. Since the *Mahatma Letters* were written in the early 1880's, at a time when the atom was still a theoretical entity (its size was not known until about 1900, and the possibility of splitting it had not even been thought of), I could surmise that the findings of science since then might provide identification of the three kingdoms preceding minerals. The mineral kingdom, of course, must be molecular, for all materials are composed of molecules. Since *atoms* combine to form molecules, they must constitute a kingdom immediately prior to molecules, and the *protons* and *electrons* which form atoms must constitute another.

There remained one further prior kingdom to identify, and for a while I called this subnuclear. It was not until quite some time later that

* Sinnett, Alfred Percy. *Mahatma Letters to A. P. Sinnett.* Transcribed and compiled by A. T. Barker. London: T. Fisher Unwin, 1923.

I realized it must be *light*. Light, itself without mass, can create protons and electrons which have mass. Light has no charge, yet the particles it creates do. Since light is without mass, it is nonphysical, of a different nature than physical particles. In fact, for the photon, a pulse of light, *time does not exist*: clocks stop at the speed of light. Thus mass and hence energy, as well as time, are born from the photon, from light, which is therefore the first kingdom, the first stage of the process that engenders the universe.

Thus I now had seven kingdoms: light, nuclear particles, atoms, molecules, plants, animals, and a seventh, as yet unnamed, of which man would be one manifestation.

The next major step occurred when I chanced upon an old zoology book, published in 1915. The author divided the animal kingdom into eight grand phyla. (The phylum is the major category of the animal kingdom.) By ignoring the horizontal difference between starfish and mollusks, I could recognize seven, rather than eight, levels of organization in the evolution of animals. This suggested that each kingdom might itself be thought of as a process, and I was able to divide the other kingdoms into seven substages and in some cases to predict categories I had not known existed (as had been done when the periodic table was introduced). The periodic table, in fact, was unequivocal in its support of the theory, for it divided all atoms into seven "periods," which show as the rows of the table itself.

This substage division made available to me information from a number of disciplines. It not only confirmed the theory, but also provided detail and feedback to refine it. What is most important, it suggested the peculiar power of the seventh stage, which I deduced from what was common to the seventh substage of prior stages: namely, *dominion*, such as DNA possesses over molecular processes in organic life.

These and other points will, of course, be covered elsewhere in the book. The whole theory is briefly outlined in Appendix I. In this Introduction I am describing only how the theory developed. The grid (see Chapter VII), or breakdown showing seven stages each divided into seven substages, was an important step because it gave backbone to the theory and made possible the principle of symmetry which I will later explain.

Perhaps the most important development was the recognition of the first kingdom as light. This took me into an investigation of quantum physics and of the developments growing out of the discovery by Planck in 1900 of the quantum of action. These revolutionized physics and revised the very basis of scientific thought. As I will endeavor to show in what follows, they provide the possibility of an entirely new view of the universe.* The older concept of a universe made up of physical particles interacting according to fixed laws is no longer tenable. It is implicit in present findings that *action* rather than matter is basic, action being understood as something essentially undefinable and nonobjective, analogous, I would add, to human decision. This is good news, for it is no longer appropriate to think of the universe as a gradually subsiding agitation of billiard balls. The universe, far from being a desert of inert particles, is a theatre of increasingly complex organization, a stage for development in which man has a definite place, and without any upper limit to his evolution.

In this drama man is at a critical point. He is more than the beasts in that he is in a different kingdom, but in this kingdom he is still not very far along. He is, in fact, at its midpoint, at a stage corresponding to that of the clam in the animal kingdom. Like the clam, he is buried in the sand with only a dim consciousness of the worlds beyond. Yet potentially he can evolve far beyond his present state; his destiny is unlimited.

* The theory we are presenting, in fact, accounts for many points that current thought still regards with discomfort and puzzlement: the uncertainty of individual photons, the curious capacity of light to anticipate its future, and the lack of identity of protons and electrons. Most importantly, it anticipates the higher orders of organization exemplified by life.

The reflexive universe

I | The fall

Cumulative nature of process

In the Introduction I have explained how I was first led to a theory of process by the importance of time, then to the idea that process must have seven stages, and from that to seven kingdoms in nature. The concept that each stage was itself a process provided additional information. Finally, the assignment of the first stage to light took me to quantum physics and to the realization that the quantum of action is fundamental.

Now consider this sequence of kingdoms:

Light
  Particles
    Atoms
      Molecules
        Plants
          Animals
            (Man)

A moment's reflection will reveal that these kingdoms or powers are not different sorts of things, like peas, apples, and oranges. The relationships which they exhibit are of another order: they are *cumulative*, they include one another, in the sense that animals include the principle of cell division first developed in plants, plants organize molecules, molecules combine atoms, atoms organize protons and electrons, and the latter are in turn convertible into photons. Each kingdom, and each power, includes what has gone before and adds a

1

contribution of its own. Each kingdom is a level of organization which depends on the one preceding.

Necessity of distinguishing levels of organization

One often hears it said, even by good scientists, that man is just a bunch of molecules, as if to deny any right he might have to a status of his own. The statement even denies him status as an animal, or as a cellular organism. This is nonsense. If one denies the animal level of organization or the cellular, one must deny the molecular, for a molecule is nothing but atoms, and the atom is nothing but protons and electrons. And lest one might think that here we have solid ground, it then appears that the electron is nothing but a "probability fog" or, if we press further, a quantum of action (a photon) that has frozen into mass.

I am not objecting to the sweeping away of all categories, as implied in the Hindu tradition with its insistence on the "ego-lessness" of things, or in the big bang theory of the universe as having been created by an initial explosion, or in the theory of its ultimate collapse into a singularity. My objection is to the singling out of molecules or of atoms for special sanction. Why stop there? It is true that molecules have, at ordinary temperatures, a relative permanence. Is it this appearance of stability that commends them, like the billiard balls, as the ideal reference that science can accept?

Man asks for bread, and science gives him a stone. He had better turn back to the old myths and get the story of the wholeness.

The fall as revealed by the discoveries of science

This brings me to one more basic idea which is important to the theory of process, an idea that is expressed in the old myths, in Plato, in almost every religion: that of a *fall*, a descent into matter, often though not always followed by an ascent back to celestial spheres and a higher state of being. This idea is rather at odds with current rational thinking, which regards itself as having outgrown such superstitious guilt-ridden notions, and as now enjoying the "enlightenment" provided by science.

However, recent developments in physics, quantum physics in particular, when properly understood, provide confirmation for the ancient notion of a fall. We may now show that it is true, in fact, that a fall occurs, for the same process by which light first precipitates or condenses into matter—losing a degree of freedom in exchange for permanence—is continued with the generation of atoms and again with their combination into molecules, so that in the grand scheme of evolution the first four stages constitute a descent from the freedom of light to the inertness that characterizes minerals (see end of this chapter).

Science has been able to discover laws of matter by restricting itself for the most part to inert objects. Galileo found the laws of falling bodies by dropping weights, not birds or moths. By concentrating on inert objects, science has made enormous progress, and there is no questioning its discoveries insofar as they are applied to their proper province, namely, to molar objects—objects containing so many particles that the agitations of individual particles cancel out and the gross object is inert.

Newton described himself as "standing on the shoulders of giants," meaning that he built on the work of his predecessors Copernicus, Kepler, and Galileo. It was he who recognized that the same law which causes an apple to fall also causes the moon to revolve around the earth.

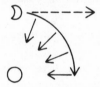

Gravity pulls the moon out of the straight line and into the circle of earth

Newton's theory of gravitation made it possible to accurately predict the motion of the planets. It gave science its *first comprehensive theory* and laid the basis of the billiard ball hypothesis, the belief that the universe could be accounted for as the motion of inert objects interacting according to exact laws. Up until fifty years ago, the expectation in physics was that this belief would apply all the way down to the ultimate constituents of matter, be they atoms or whatever.

The existence of atoms was first hypothesized by science to account for what were called combining ratios. Soon after 1800 it was found that whenever different materials were combined to create a third—as oxygen with hydrogen to form water, or H_2O—they did so in fixed ratios: 8 grams of oxygen to one of hydrogen. This fact could be explained by supposing that all materials are composed of atoms whose relative weights are in fixed ratio. This hypothesis proved quite satisfactory and could be extended to all materials, despite the fact that the absolute size of the atom was unknown.

Not until 1900 was it possible to actually count the atoms in, say, a gram of hydrogen. The number was found to be huge: 6.02 times 10^{23} atoms in one gram of hydrogen or in 16 grams of oxygen. (The oxygen atom is sixteen times the weight of the hydrogen atom and, of course, combines with two hydrogen atoms in H_2O, making the combining ratio 8:1.)

But no sooner had the atom been at last tracked down than there came a disconcerting discovery: the "indivisible" atom was made of parts! And these parts, of equal and opposite electrical charge, were unequal in weight: the proton is eighteen hundred times heavier than the electron.

Then came the question: how big is the proton? In an ingenious experiment, Rutherford bombarded atoms with alpha particles (helium atoms with electrons stripped off), and showed that the proton could be no larger than 10^{-13} centimeter—about 1/100,000 part of the whole atom! This was another surprise, for it emphasized still further how utterly minute were the nuclear particles of the kingdom prior to that of atoms.*

But the fact that atoms had turned out to be divisible, despite the indivisibility they were formerly thought to possess, provided a beautiful simplification, for now it was possible to account for the ninety-two *different* atoms as made up of only two ingredients, the electron and the proton! And it was also found that the chemical and other properties of atoms depended solely on the *number* of electron–proton pairs. The number of such pairs is called the atomic number. Each kind of atom

---

* I neglect the transitory particles—mesons, etc.—because of their brief life spans, about one-billionth of a second.

has thus a different number assigned it, and every possible number of proton–electron pairs up to ninety-two is a different kind of atom.

A more eloquently simple way of obtaining the complexity of matter, with its ninety-two kinds of atom and its countless kinds of molecule, cannot be conceived. It confirms in yet another way the insight of Pythagoras that all is number.

So the discovery that atoms could be divided into more elementary constituents was a triumph of rationality. I should say another triumph, for meanwhile the precise frequencies of the light radiated or absorbed by atoms were accounted for in Pythagorean fashion as the ratios of whole numbers. Simple, elegant, rational, the entire behavior and structure of atoms could be figured down to the last decimal point. The rationality of science enjoyed an undreamed-of triumph. The billiard ball hypothesis was apparently confirmed.

Uncertainty enters the picture

But then, from an unexpected quarter, came a shattering blow. So far, in working out the details of atomic theory, it had been assumed that the ultimate units, the proton and electron, were the same *kind* of thing as inert objects—that they were just very small billiard balls, but in other respects like ordinary objects such as grains of sand or specks of dust. However, this assumption had overlooked an important consideration: how are you going to *see* these small particles? The dénouement was like Portia's caution to Shylock in *The Merchant of Venice*, that he may have his pound of flesh, but if he sheds one drop of blood his life is forfeit. In the modern version, Heisenberg has the role of Portia. He pointed out that for the presumption of predictability to be carried out, it is necessary *to observe* the position and momentum of this tiny particle. And to observe it we must throw light on it—right? But the wavelength of ordinary light (10^{-5} centimeter) is a million times greater than the diameter of the particle you are observing; so it won't provide any accuracy (a difficulty that limits the resolving power of optical microscopes to three-thousand-fold).

Very well, let's use shorter-wavelength light. But since this shorter-wavelength light (x-rays) is in the form of photons which carry enormous energy, their impact upon the particle will knock it out of the picture.

Hence the predicament: the observation of position and momentum required for prediction *cannot be carried out*, and the determinism we were led to expect at the level of molar aggregates (objects made of billions of particles) does not hold for elementary particles; their position and momentum are indeterminate.

Reader, do you sense what's going on? Or do you feel perhaps that some probe finer than light can be found which will make possible the observation without the disturbance? Many scientists have felt the same, for one does not like to give up one's faith that the universe can be known, or in other words, that it is *objective*. But as Heisenberg was at pains to point out, we are up against not merely the physical limitation of an instrument which a better instrument might circumvent. We encounter here a principle which imposes a theoretical limit on the accuracy of our knowledge about individual particles.

For however we vary our probe, the uncertainty is the same; we can measure position with accuracy only at the expense of disturbing the momentum, and we can measure momentum only at the expense of disturbing the position; the product of the two uncertainties is a constant.

Implicit in this dilemma is an important philosophical principle: the observer can know the universe only by interacting with it, and this interaction requires more energy as the accuracy of observation increases. A remarkable finding, for it makes zero as unattainable as infinity!

But our concern in the present context is with the entry of uncertainty into the picture. The predictability which held for molar aggregates (stones and other inert objects) is not possible with ultimate particles. It might be objected that the uncertainty is *introduced* by the act of observing them (epistemological). But it has been found that the particles carry or possess the uncertainty anyway (ontological.)* There

---

* That the electron "possesses" an innate uncertainty is implicit from the definition of the fine structure constant. See p. 35, "level II." See also Northrop's introduction to Heisenberg, wherein Northrop emphasizes that the uncertainty of the electron is *ontological,* not just epistemological. Heisenberg, W. *Physics and Philosophy.* New York: Harper and Row, 1958.

is only a probability that they are in a given place. The electron is now referred to as a "probability fog."

Another attribute of these fundamental particles, protons and electrons, is that they have no identity. It is impossible to know whether an electron leaving an atom is the "same" electron as the one which entered it. In order to carry on with the task of science, to discover laws and to predict, the physicists have given up attempting to deal with *individual* particles. Like insurance companies, they must use statistics. They even have different kinds of statistics for different kinds of particles: Boltzmann statistics for molecules that can be distinguished; Fermi–Dirac statistics for electrons that cannot.

I will not dwell on this here except to stress that our story has a moral: the world of fundamental particles is quite different from that of predictable billiard balls. From the point of view of predictability, it is like that of human beings. Its creatures have a life of their own. Predictability here is similar to that of insurance tables, Gallup polls, and market surveys: it does not apply to *individuals*. The individual particle does not obey laws.

The fall into determinism

Why is this important? Because here in science we discover that *the evolution of matter itself is a "fall."* By fall we mean loss of freedom, increase of constraint. This occurs in steps: first the condensing of the original energy of the photon into mass to form a charged particle, then the joining of opposite charges to constitute a neutral atom, then of atoms to form molecules, and finally the compaction of molecules into inert objects. In this declension the original freedom is lost: the free motion of one particle is canceled by the free motion of another, so that where billions of particles are compacted, there results an *inert object*, which, having no self motion, responds to exact laws.

It is these inert objects that are the basis for the so-called universal laws. It is these objects and only these that do not move unless pushed.

We must therefore interpret the generation of matter—of the universe itself—*as a fall into determinism.*

Beyond determinism

But this is not the end of the story. There are higher forms of organization. Not all molecules are compacted into inert objects. There are some which are *organized* into living creatures: plants, animals, human beings. Can we read these higher forms as an ascent— a "return"—to freedom?

I believe we can, and I will introduce this image of an "arc" as a basic postulate. It is one of the concepts I will use in the theory that I am setting forth. I could also say that I found it borne out, for in examining the kingdoms it became evident to me that there was a fall followed by an ascent: the first and final kingdoms being most free and the ones in the middle the most determined.

It is almost self-evident that this should be the case, for process, as the dictionary defines it, consists of steps taken to reach an end "as in the process of making steel." Therefore *any process projects a goal and goes through means* to attain it. Such means are necessarily determinate or predictable. If our presumed means do not function as we expect, we find other ways to achieve the end. If our car breaks down, or in any way becomes unpredictable, we resort to other means of transport. (See discussion of the work of an inventor in the Introduction.)

The reader might ask if the increase in determinism which science has discovered in the generation of matter is properly interpreted as a fall of the same kind as described in myth.

In answer, let us cite the Egyptian myth which begins when Osiris is trapped by Set in a coffin fashioned for the purpose. The coffin then floats down the Nile and comes to rest at the base of a tamarisk tree, which grows up around it, *enclosing* it. Finally, Osiris is dismembered by Set and the pieces strewn in the marsh. (The myth continues with Isis, who reassembles the pieces and conceives from the corpse the infant Horus, who becomes the hero who conquers Set.)

This myth, with its description of increasing constraint, first by the coffin and then by the enclosing tamarisk tree, presents the fall in terms of loss of freedom. The next stage, the dismemberment of Osiris into pieces, involving total loss of any possible initiative or self-motion, represents a state of complete inertness or determinism.

The reader may object that the story of Osiris, like the third chapter of Genesis, refers to the fall of man, not to the generation of matter. But I insist that both myths nevertheless describe process and its necessary descent into means to attain its end. *The advanced stages of organization which constitute life must have determinate matter to work with:* Osiris must be dismembered in order that Horus may be born. This is the *universal* pattern of process, experience, development.

The arc

We may therefore depict this descent and ascent through which process develops as an arc:

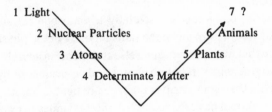

1 Light
2 Nuclear Particles
3 Atoms
4 Determinate Matter
5 Plants
6 Animals
7 ?

The mere idea of a fall gives only the overall picture, but this arc arrangement yields clues to why process is what it is. By visualizing the seven stages as occurring on four levels, each having a distinct character, we can uncover further details which will be developed in succeeding chapters.

II | Light as purposive

The enigma of light

The heart of our story, like the beginning of creation, lies in the nature of light. Here we are confronted with mystery, and I say this not because of the perplexities of physics, but in view of what is the essential nature of light. Light, because it is primary, must be unqualified—impossible to describe—because it is antecedent to the contrasts necessary to description.

While the foregoing is essentially a philosophical statement, the physicists would have much the same report. For the physicist, light is unique in that, unlike everything else that exists in actuality, it has no mass (no rest mass). It has no charge and, as evidenced by the finding of relativity that clocks stop at the speed of light, it has no time.* While light in a vacuum has a "velocity" of 186,000 miles per second, this velocity is not motion in the ordinary sense since it can have no other value. Objects can be at rest or move at a variety of speeds. Light, on the other hand, has but one speed (in any given medium) and cannot be at rest. Even space is a meaningless concept for light, since the passage of light through space is accomplished without any loss of energy whatever.

Light involves us in a special kind of difficulty, the difficulty of knowing about that which provides our knowledge of other things. We might imagine a painter who wanted to paint the paintbrush, a problem I encounter when I want to repair my glasses: I cannot see without them; and light, by which we see, cannot be seen.

* The space–time path of light has zero length.

10

This sort of Zen paradox is not appreciated by the scientist, who likes to think of light as "just another kind of particle." This interpretation does not stand up under examination, for to call that which is outside of space and time, and which has no rest mass, "just another kind of particle" is a placebo for materialists rather than a correct description (see p. 6).

Light is not an objective thing that can be investigated as can an ordinary object. Even a tiny snow crystal, before it melts, can be photographed or seen by more than one person. But a photon, the ultimate unit of light,* *can be seen only once*: its detection is its annihilation. Light is not seen; it is seeing. Even when a photon is partially annihilated, as in scattering of photons by electrons, what remains is not part of the old photon, but a new photon of lower frequency, going in a different direction.

An ordinary object can be thought of as a carrier of momentum, or energy, which it can impart to another object. A hammer striking a nail exerts a force which drives the nail; a bowling ball conveys energy which knocks over the pins. In both cases, the hammer and the bowling ball remain after the work is done. With light, however, its transport of energy from one point to another leaves no residue. *Light is pure action*, unattached to any object, like the smile without the cat.

This light energy is everywhere, filling the room, filling all space, connecting everything with everything else. It includes much more than the light we see by, for *all exchange of energy* between atoms and molecules is some form of what used to be called electromagnetic energy, which extends over a vast spectrum and would be better named interaction. Visible light covers just one octave in that spectrum.

We have said that light is unqualified, by which we mean that the usual ways in which we distinguish one object from another are not available for its description. However, light does have frequency and wavelength. The frequency is the velocity of light divided by wavelength. The visible-light wavelengths are recognizable as color, red light having long wavelength and low frequency, and violet light shorter wavelength and higher frequency. Over the range from red to violet, the frequency of light doubles (one octave). The entire range of wavelengths of light, between that commensurate with the diameter of

* The neutrino, if it exists, is no less insubstantial than light.

the proton (10^{-13} centimeter) and the long waves used in transatlantic radio, covers at least sixty octaves.

The electromagnetic spectrum

Frequency (cycles per second)		Wavelength (centimeters)
10^{23} cps	Creation of proton	10^{-13} cm
	Gamma rays	10^{-11} cm
	X-rays	10^{-8} cm
	Ultraviolet	10^{-6} cm
	Visible	10^{-5} cm
10^{14} cps	Infrared	10^{-4} cm
	Heat	10^{-2} cm
	Radar	1 cm
	UHF (television)	10^{2} cm
10^{6} cps	100-meter radio broadcast	10^{4} cm
	20,000-meter radio	10^{6} cm

Early explanations of light

Newton first observed the decomposition of white light into colors when he caused a narrow beam of sunlight in a darkened room to pass through a prism and fall on a screen. He considered light to be corpuscular in nature, i.e., to consist of tiny particles. Observing the rings, or interference fringes, visible when a curved glass surface rests on a flat one, he surmised that the corpuscles must have a periodic nature, but he denied that they were waves.

Huygens proposed the wave theory in 1690. After commenting on the extraordinary speed of light and the fact that its rays interpenetrate without impeding one another, he urged that light could not be due to transmission of matter, as for instance a projectile, but must, rather, resemble the transmission of sound through the air. This motion, he concluded, must be gradual and spread in spherical waves, similar to waves in water when a stone is thrown into it.

So great was Newton's authority, however, that the wave theory did not gain general acceptance until 1800. Let us note that in both theories there is *a dependence on something material*, either the corpuscles or the medium in which the waves are conceived to be transmitted.

The 19th century ushered in the discovery of electricity. Galvani's experiments with frogs' amputated legs led Volta to his electric battery. The battery led to electric currents in wires and to Ampère's formulation of the laws which relate electricity to forces. Finally, the brilliant experiments of Faraday brought all electrical phenomena together under his theory of a field in which "lines of force" were invoked to account for action at a distance.

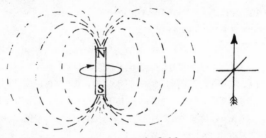

Diagram of a magnetic field

The magnetic field, as already mentioned in the Introduction, is toroidal in shape.

It was therefore an extraordinary achievement when, in 1873, James Clerk Maxwell was able to bring all of Faraday's theory of fields and the wave theory of light under one general mathematical formulation. This formulation could explain light, electricity, and magnetism; it also predicted radio waves, which were soon confirmed by Hertz.

The wave theory of light required the existence of an all-pervading ether as the "agency" for the transmission of electromagnetic forces. The ether, while it solved the vexing problem of action at a distance, created new problems. Force was explained as a "strain" in the ether, but in order for the strain to travel at the speed of light, it would be necessary for the ether to possess a rigidity *millions* of times greater than the rigidity of steel (according to de Broglie, a modern physicist, its rigidity would have to be infinite). The ether seems a remarkable example of how far people will go to shore up and thus obscure faulty thinking rather than abandon a mistaken premise, i.e., materialism.

But this absurdity did not dislodge the theory of an all-pervading ether. It was only when, in the 1880's, the experiments of Michelson and Morley compared the speed of light perpendicular and parallel to the earth's motion, and showed no difference—and therefore no way of detecting motion with respect to it—that the ether hypothesis began to be questioned.

The quantum theory

There was one fact which the wave theory of light could not explain: the phenomenon known as the photoelectric effect. When light falls on a metal plate, it knocks out electrons, but the velocity of the emerging electrons does not depend on the intensity of the light but on its frequency. The wave theory would predict the velocity to depend on intensity. (A similar problem was to explain photography. No matter how faint the light, even from a distant nebula, it still has enough energy to separate the silver bromide molecule and release the free silver which darkens the negative.)

To explain the photoelectric effect, Einstein in 1905 called upon the theory that had been proposed by Planck in 1900, that *light is transmitted in whole units*, or *quanta of action*. These quanta of action, also called photons, are the dynamic counterpart* of the more familiar atoms of matter, and are transmitted without attenuation. This theory held that the photon contains an amount of energy (E) proportional to its frequency (F).

$$E = hF \ (h = \text{Planck's constant})$$

Or, since frequency is the inverse of time, $E \times T = h$. (A wave with a frequency of 60 cycles per second has a period of time of $1/60$ second.)

Because the energy of the photon (determined by its frequency) is not diminished with distance, the photoelectric effect is accounted for.

---

* Eddington described photons as "atoms of field action."

But the quantum theory had to wait a long time for its acceptance. When the evidence came in, it was overwhelming.

For example, the discovery of Heisenberg, to which we have already referred, that the electron cannot be observed without being disturbed, was found to require Planck's theory. For, in this case, the product of the uncertainty of position times the uncertainty of momentum *is also a quantum of action* equal to Planck's constant.

These and other discoveries fell into place as part of the revolutionary new physics known as *quantum physics*, which established:

1. That light is radiated in whole units which do not dissipate their energy on the way to their targets (quanta of action).
2. That all energy exchange at the atomic and even the molecular level is in terms of quanta of action (light).
3. That action, like matter, comes in discrete whole units which cannot be divided.

What led Planck to this important discovery? As the story is generally told, Planck was trying to account for the discrepancy between theory and fact in the explanation of how radiation varies with temperature. According to the theory then in vogue, the radiation energy of a heated object should be equally distributed for all frequencies. Imagine a box containing radiation of all possible frequencies or wavelengths:

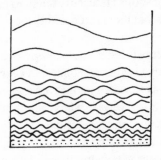

Since there *is a lower limit* to the frequency in a given size of box (its wavelength can be at most twice the size of the box), and there is no upper limit, classical theory predicted that the energy would be absorbed by the higher frequencies. The result would be what was known by the engaging title: "the ultraviolet catastrophe."

But this is definitely not the case. The radiation *has* an upper limit beyond which it fades out:

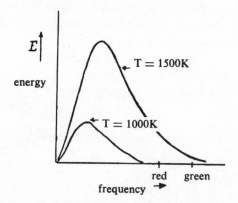

Thus an object, as it is heated, first gets red hot, then white hot, moving toward the blue only as the temperature increases. It requires more energy to radiate at the higher frequency.

It was to explain this unexpected distribution that Planck assumed the energy is radiated in packets whose energy is proportional to the frequency. That is, Energy × Time = Constant. This is Planck's constant and is the unit of *action*.

Thus it required increasingly more energy to supply the high-frequency vibrations (like more talent to fill executive positions, so that it remains true that there is room at the top despite the housing shortage).

That greater energy is required for *shorter* wavelengths of light (that are higher frequency) is not what one would expect on the basis of sound and water waves, which require more energy the *longer* they are. This is only one of many ways in which quantum theory has acted as a check and a correction on rational intellect. Rational process is a splendid assistant but a poor guide in almost any field, especially in the fundamental questions where the normal polarities fade out or reverse.

The principle of least action

The difficult question is: what is *action*? This will become increasingly important as we proceed. Curiously, the notion of light as action was one to emerge quite early. It was observed in the 17th century that sunset occurred a little later than it would if light followed a straight line: light as it enters the atmosphere follows a curved path. This phenomenon is explained as due to the fact that the speed of light is reduced by the atmosphere.

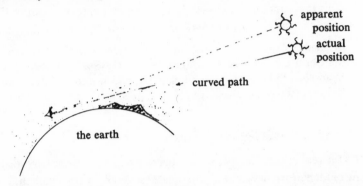

What is remarkable is that the path followed by the light through the layers of atmosphere is precisely that which gets it to its destination in the shortest possible time. In driving from a point in the city to a point in the country, we can reduce the total *time* if we shorten the time spent in the city, even at the expense of going a longer distance. Fermat, the famous 17th-century mathematician, was the first to solve this problem of the path for the minimum time. Yet light, going from a denser to a

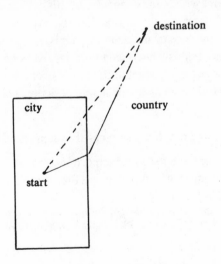

rarer medium, follows just this path. As Planck himself said of the phenomenon:

> Thus, the photons which constitute a ray of light behave like intelligent human beings: Out of all possible curves they always select the one which will take them most quickly to their goal.*

This law, that light always follows the path taking the shortest time, is known as the *principle of least action*. According to Planck again:

> [It] made its discoverer Leibniz and soon after him also his follower Maupertuis, so boundlessly enthusiastic, for these scientists believed themselves to have found in it a tangible evidence for an ubiquitous higher reason ruling all nature.

Consideration of light as purposive

As the reader is probably aware, the notion of purpose or teleology is forbidden in science, among biologists especially, who, while they must be strongly tempted to invoke it at every turn, avoid it as a reformed alcoholic avoids a drink. Physicists avoid it because their problems don't require it.

Yet we find one of the greatest physicists saying that:

> . . . the historical development of theoretic research in physics had led in a remarkable way to a formulation of the principle of physical causality which possesses an explicitly teleological character.**

But I do not wish to make an issue of this question of teleology here. Let us simply note one thing: that there is only one exception to the exclusion of purpose from science, and this exception is light, which these several scientists have seen fit to regard as having a purposive behavior. Let us also note that the purposiveness is associated with that aspect of light known as the principle of action (or least action).

Importance of Planck's discovery that action comes in wholes

What did Planck add to this principle of action that was not already present in the ideas of Leibniz? It was the notion that action comes in

* Planck, Max. *Scientific Autobiography and Other Papers* (p. 178). Translated by Frank Gaynor. New York: Philosophical Library, 1949; Greenwood Press, 1968 (reprint).
** *Ibid.*, p. 80.

quanta or *wholes*, and that this unit is constant. Note that despite the tendency to refer to energy as quantized—a habit which even good physicists are given to—it is not energy but *action* that comes in wholes.

$$\text{Action} = E \times T \text{ (Energy} \times \text{Time)} = \text{Constant } (h)$$

Action is constant, energy is proportional to frequency. (T is the time of one cycle.)

So far, except for the reference to purpose, I have kept within the bounds of accepted science. Now I would like to go further to track down this notion of purpose which Planck, and before him Leibniz, felt was indicated by the principle of least action.

As we have noted, purpose is barred from science. As Bacon said: "Purpose like a virgin consecrated to God is (for science) barren."

But as Whitehead pointed out in his *Function of Reason*:* "Scientists, animated by the purpose of proving they are purposeless, constitute an interesting subject for study." As Whitehead went on to say, we must distinguish "between the authority of science in the determination of its own methodology and the authority of science in the determination of the ultimate categories of explanation." Whitehead obviously wants to include purpose as an ultimate category of explanation.

How may we include purpose in cosmology (the ultimate categories of explanation) while still excluding it from the methodology of science?

We know that science builds its entire edifice on three measures: mass, length, and time, and their combination, and all scientific formulation can be expressed in these terms. Clearly, there is no evidence of purpose in any of these: it is not in mass, nor in length, nor in time.

The only suspicion of it, as we noted, occurs in the formula for action. Action has the measure formula ML^2/T. This *combines* mass (M), length (L), and time (T). Is it possible that there is something present in the whole that is not in the parts?

This is clearly the case here. Consider any device made of parts, say a bottle and its cork or a flashlight and its bulb. Is it possible to find the

---

* Whitehead, Alfred North. *The Function of Reason.* Princeton: Princeton University Press, 1929.

function of the device in the parts? Surely, no. Only when the device is put together can it express its function and its purpose, something its parts alone could never do.

It was Planck's epoch-making discovery that action *comes in wholes*, a discovery which in retrospect we can see to be true of human actions. We cannot have 1½ or 1.42 actions. We cannot decide to get up, vote, jump out the window, call a friend, speak, or *do* anything one-and-a-half times. *Wholeness* is inherent in the nature of action, of decision, of purposive activity. Planck's discovery about light touches home: it is true of our own actions. But we didn't really know this until the physicists had made this a principle.

Light as first cause

Perhaps I should let it go at that. We are already pressing the mind beyond its limits. Nevertheless, let's go ahead and see what happens. Since purpose is in the whole and not in the parts, the whole must be greater than the parts. How can we account for this? Because the whole *cannot function when divided*. It follows that function is that aspect or "cause" which is not in the parts and which science cannot deal with, because science deals with mass, length, and time, which are parts. This leads to a basic cosmological postulate: *the parts are derived from the whole*, and not the whole from the parts. In other words, the whole exists *before* the parts (see pp. 149–150).

We can now close our argument, for in showing that the parts arise from the whole we provide confirmation for *light as first cause*:

Light = quanta of action = wholes = first cause

An additional consideration that confirms the fundamental nature of action is that actions are *unqualified*. While mass is measured in grams, length in meters, and time in seconds, quanta of action are *counted* with no necessity of specifying the kind of unit. This implies their fundamental nature; actions *precede* measure, they are prior to the analysis which yields grams, meters, and seconds.

It might be objected that action has the measure formula ML^2/T and hence cannot be dimensionless. The answer is that, though action has the dimension ML^2/T, we are taking the position that this particular combination of dimensions (known as action) is the *whole* from which time, mass, and length are derived. The reasons are as follows:

1. Action comes in irreducible *quanta* or units.
2. These units are of constant, i.e., *invariant*, size.
3. They are counted, not measured.
4. Because indeterminate, they constitute the end point in the chain of causation and are therefore *first cause*.

Additional information

Our history of concepts of light in this chapter culminated in Planck's discovery that light is radiated in quanta, or whole units of action, h, whose measure formula is ML^2/T (or $2\pi ML^2/T$).

In the interest of completeness, we should here note that the quantum of action has yet another alias—*angular momentum*. For angular momentum and action have the same measure formula: ML^2/T. This need not make things more difficult, for angular momentum is a clearly defined concept and more easily understood than is the concept of action or uncertainty, and hence can aid in our understanding of action.

But what is angular momentum? It is the momentum of a spinning weight. *Linear* momentum, which does not involve spin, is the momentum of a hammer or of a car hitting a telegraph pole and is a product of mass times velocity, i.e., MV or ML/T since V, or velocity, is length divided by time. But angular momentum is linear momentum times the radius around which it spins, $ML/T \times L = ML^2/T$.

Linear momentum: MV

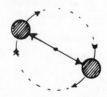

Angular momentum: $MV \times L$ or ML^2/T

Thus a weight spinning around an axis or two weights spinning around a center possess angular momentum. If we imagine that the wire holding the weights breaks, then we can explain the uncertainty (of the quantum of action) as our uncertainty about the direction in which the released weight would fly.

Thus angular momentum "packages" momentum so that it no longer requires, as does linear momentum, a velocity of translation to exist. For example, a spinning flywheel has angular momentum and hence can store energy, even though the flywheel remains in one place. Yet the flywheel can be transported and thus convey its stored energy.

This helps answer the long-standing question of action at a distance. For according to this theory, light contains energy in its angular momentum, and can thus transport this energy from one point to another. Newton's corpuscular theory could not explain the transmission of different amounts of energy, because the corpuscles, which must move at the speed of light, all have the same speed, and could thus not account for the great difference in light energy. (See electromagnetic spectrum, p. 12.) Alternatively, the wave theory could explain the transmission of energy only by postulating an infinitely rigid ether.

Moreover, Planck's constant—the quantum of action—can contain *any amount* of energy. This may be understood by the example of a skater or ballet dancer who, giving his body a whirl with arms extended, makes himself spin faster by drawing his arms in close to his body. In the case of the photon, there is no smallest size, and—as the radius is decreased—the spin gets faster and faster (the frequency of the photon increases). This faster spin has more energy, and there is no upper limit.

But where does this energy (for the skater) come from? It is produced by the skater in pulling in his arms against centrifugal force—and would be greater if the skater held weights in his hands. Of course, with the skater there is a limit to the rate of spin or the energy he can store in this way because he can pull his arms in only so far—but the photon, having no bulk, can shrink any amount. It follows that a single photon can store unlimited energy by getting smaller!

This is one of the most surprising findings of quantum physics—that the *smaller* the photon, the *more* energy it contains. It is a reminder of the great difference between the world of light and the world of matter.

World of light	World of matter
No time, space, charge, or mass.	Time, space, charge, and mass.
Energy increases as size (wavelength) is reduced.	Energy (mass) decreases as size is reduced.
Constant velocity; no rest.	Any velocity less than the speed of light.

Summary

This chapter has been devoted to establishing a connection between light, purpose, and first cause.

We have shown how science, in regard to its understanding of light, has been put through the hoop, having been forced to reverse itself twice: from "corpuscles" to wave theory to quanta of action. We may therefore appreciate that light is not "just another particle," nor is it describable as most things are in terms of something else. It is the base element.

We have given some account of Planck's discovery of the quantum of action, or photon, as the ultimate unit of light, and have indicated the connection of action with purpose. We have noted that purpose, inasmuch as it is decision, comes in wholes, and that there is a necessary connection, in that when wholeness is divided, purpose or function ceases. We conclude that light is the unitary purposive principle which engenders the universe, and that it has the nature of first cause.

III | More on light

Prescientific concepts of light

I would like in this chapter to go back to some opinions on light that were held long before the emergence of modern science. This is to the period around A.D. 350, when philosophers were endeavoring to rationalize the account of creation in Genesis, and before the dogma of the church had set in and discouraged speculation.

According to I. P. Sheldon Williams, St. Basil of Caesarea, writing in about A.D. 370, distinguished between the intelligible and the sensible world.* The intelligible world was outside time. The sensible world shared with it "an intelligible matter which Basil identifies with the light which illuminates the material world and is therefore the common ground of the whole universe intelligible and sensible."

> It follows [says St. Basil] that light is of a more general nature than time, for time is found only in the sensible world. . . . Because light is not limited to time it was universally distributed at the moment of its creation, as it fills the whole room in which a lamp is kindled. Between the intelligible and the sensible worlds the firmament acts as a barrier, of which the solidity implied by its name is such that light may pass through it (but in a diluted form), but time cannot break out into the world above.

Now this is a most remarkable statement because it recognizes that time does not exist in the world of light, a fact known to science only since the advent of relativity!

The recognition of a basic division between an "intelligible" and a

---

* *Hexameron* V 2, 5:40c–41a, 7:45a. Quoted in Williams, I. P. Sheldon. *The Cambridge History of Later Greek and Early Medieval Philosophy*. Edited by A. H. Armstrong. London: Cambridge University Press, 1967.

"sensible" world is common to all prescientific thought, and the belief in an intelligible world began its decline only when comprehensive theories, like that of universal gravitation, apparently made it possible to explain the phenomenal world in mechanical, or more correctly, reductive, terms. While Newton believed that the regularities of planetary motion were evidence for the divine, Laplace, his successor, did not. In answer to Napoleon, who had commented that Laplace had written his *Celestial Mechanics* with no reference to its Maker, Laplace replied: "Sire, I have no need of that hypothesis."

This has been the trend in the philosophy of science ever since. But a careful examination of the *findings* of science, as distinct from its *philosophy*, will disclose that at the true frontier of science this trend has reversed: as we pointed out in the preceding chapter, the more deeply the physicist prods into the foundations, the less certainty he finds. It is scientism and not science that still clings to the old determinism. It is the behaviorist, the sociologist, who "any day now" will discover the "laws" of human behavior. The layman too has endorsed the credo of determinism, perhaps because he feels that it is scientific, but also because it gives him security, it exorcises the demons of change.

The physicist, who was originally responsible for the doctrine (which Blake called "Newton's sleep"), has been awakened by his own experiments to the recognition that determinism as a doctrine is untenable. First alerted by Heisenberg's teaching that the fundamental particle cannot be observed without disturbing it, he has found uncertainty the rule rather than the exception, and he knows that he can predict only in the statistical sense; he cannot predict for individuals.

What is remarkable is that this fundamental turnabout has not been more widely disseminated. Eddington discussed the change from determinism in his delightful fable of "Canticles." The story goes that a scholar mistakes "Canticles" for a person, about whom he writes papers and to whom he attributes various works. After some years the scholar discovers that the word means "songs," and tries to correct the mistake he has made, only to find that other scholars have adopted his brain child and resent his effort to do away with their idol. They attack him with the criticism that he has no proof that "Canticles," the person, did not exist.

In the same way, determinism, invented on the basis of a

misunderstanding, and now known to be false in principle, has been made policy, and its adherents challenge others to disprove it. Robert Oppenheimer observed in 1956:

> ... the worst of all possible misunderstandings would occur if psychology should be influenced to model itself after a physics which is not there any more, which has been quite outdated.*

Nonetheless, this has happened, and the pundits with the false dogma are making the most noise.

We should not underrate the difficulty involved. The determinism of the 19th century made a deep impact on our thinking, not because it was materialistic, but because it gave sanction to the rational mind to take over the functions which should be performed by other faculties. I am saying that it is the rationality of determinism that makes it dangerous, usurping not only the intuition of the higher mind, but practical common sense.

This is an old saying (practically a major thesis) in Zen philosophy: "The mind is the slayer of the real." There are other "enemies" that man has to conquer, such as his physical appetites, but the lusts of the flesh are less important today than the lusts of the mind, seducing leaders and the public alike with brain trusts, expertise, computers, the information explosion, and other forms of reliance on rational process.

The pertinence of this recital of what may be self-evident is that in the discovery of the true nature of light we have a reference, a guide which can help us set cosmology to rights with first things first. And now I am going to have to say something that may seem heretical to scientists and theologians alike.

This is that quantum physics, when understood, can provide the compass that in former times required revealed religion: it can tell us of first cause.

Let me go back to St. Basil. Here we have a Greek philosopher who makes far more profound and correct statements than did the philosophers who lived at the height of Greek civilization (4th and 5th centuries B.C.). Compare St. Basil with Aristotle. Impressive as was Aristotle's encyclopedic mind, he said nothing that would anticipate the

* Whitmont, Edward C. *The Symbolic Quest.* New York: G. P. Putnam's Sons, 1969.

revolutionary discoveries of modern science that were to reverse reason. Aristotle did not say that time does not exist in the world of light, nor could reason reach such a strange idea. How did St. Basil reach it? *By reasoning about revealed scripture.* This brings out the function of revealed religion: it supplies what reason *per se* cannot. We could not reason that light was the first thing to come into existence. Reason begins at the other end, with objects, and proceeds to divide the object much as did Democritus with his atoms. Reason has to start with something; it cannot admit first cause. Try it on yourself; you will always find yourself asking what was the cause of the first cause.

Genesis tells us that first cause is fiat: Let there be light! What is "fiat"? It is *action* (decision). So when Planck discovers that light is a quantum of action, he is discovering "fiat," the same fiat described in Genesis, the light that was created on the first day. But, you say, God is the cause behind the first cause.* The question of whether God (action) is different from light (action) is a mystery that I cannot resolve. I suspect misplaced concreteness.

I'm not implying that revealed religion is necessarily correct. As St. Gregory, a contemporary of St. Basil and an equally pious philosopher, said (as paraphrased by Sheldon Williams):

> The Language of Scripture does not reveal the truth directly but a half concealed version of it. This is because the sensible world from which it draws its illustration is an imperfect copy of the intelligible.**

One would think St. Gregory had been reading Eddington, who says:

> The physicist draws up an elaborate plan of the atom and then proceeds critically to erase each detail in turn. What is left is the atom of modern physics.***

In other words, we can trust neither reason nor revealed religion alone; we need both. This does not mean that all reasoned interpretations of scripture are correct, quite the contrary. But when

---

* We could also say that there is no reason for light not to have an antecedent. For light to be first cause means only that its antecedents do not imply the result. See discussion of *the turn* at the opening of Chapter V.

** *Ibid.*

*** Eddington, Arthur S. *New Pathways in Science* (p. 259). London: Cambridge University Press, 1947; Ann Arbor: University of Michigan Press, 1959.

we can find so close an equivalence as exists between the prerational accounts of light as the beginning of things on the one hand and modern physics on the other, then I think it is at last possible to begin to correctly interpret the symbolic statements of revealed religion.

Here we must be careful not to go astray, as has so often been done, and it is to guard against this possibility that I invoke quantum physics. Not all physics, which is often as much a slave to rationalism as is philosophy. Recall the physicist who insisted that the atom "could not be divided because the word atom means indivisible." Or recall the vicissitudes of the theory of light, which went from corpuscles, to waves in an ether, to quanta of action. Such a history implies that something was forcing the concept of light to grow in a way that rationalism could not have provided. This factor is, of course, experimental evidence, but it had to manifest through particular facts, such as the discrepancy in how radiation varies with temperature, which *forced* Planck to invent a new theory.

Perhaps the reader will think, "Well, physics is already the revealed religion of the present time—you are just creating a new sect." But it is quite to the contrary: I'm attempting to show that the hard-won truths of physics, those that have forced the most radical departures, are equivalent to what the best earlier thinkers discovered in revealed religion.

Thus, rather than creating another sect, I'm asking for an examination in depth, which I affirm can effect not only a confirmation of what was formerly faith, but a synthesis of science, which today is becoming so severely fragmented that it is losing all meaning and would stand to benefit as much as would religion.

So much, then, for the overall picture. We have dealt in the last two chapters with the many-faceted problem of light, all aspects of which point one way: to the ultimate centrality or primacy of light as the origin of everything. And by everything I mean not only matter, which light creates by its condensation from photons and which it changes by the interaction of photons with atoms and molecules, but also what St. Basil called the *intelligible* world, the eternal now of consciousness. Perhaps it will be discovered that tachyons, those supposed entities which travel faster than light, may comprise the intelligible world. As to that, I will not attempt to say. I will be content if I can carry the reader to the border of this kingdom in which, paradoxically, he already resides.

Preview of later chapters

In the chapters that follow we will trace evolution through the kingdoms of nature. We will find that as the entities of nature develop into higher and higher orders of organization—from particles to atoms, from atoms to molecules, thence to cells, organisms, and animals—they are invariably associated with an intrinsic activity. This intrinsic activity is the positive factor behind evolution and for two important reasons cannot be dismissed. One is theoretical, that a universe of inert particles (billiard balls) could not in itself construct the highly ordered and extremely complex entities which we call living creatures; and the other is that quantum physics has discovered in the quantum of action just such an innate activity.

The quantum of action, as it occurs, for example, in visible light, involves an extremely small amount of energy as compared with the energy we exert to walk about or drive an automobile, but this comparison is misleading. Just as the energy required to *steer* an automobile is minute compared with the energy required to *propel* it, so too the energy required in *deciding* to move our own body is minute compared with the energy of actually moving. Our body is, in fact, an elaborate mechanism in which muscles are controlled by nerves, and nerves by minute electrical changes such as are involved in changing the bond in a molecule. Changing the bond in a molecule is just what the quantum of action *can* effect, but the difference between the creation of a starch molecule from carbon dioxide and water (which is accomplished in chlorophyll by a quantum of light) and human decision is so great that we need to trace with care how nature has evolved from the chlorophyll molecule to creatures of the complexity of animals and man.

In this evolution the quantum of action will be found to have a central role. While it is itself invisible and unknowable (one of its other names is the quantum of uncertainty), it persists through the whole chain of being and "causes" the progressive movement of evolution, much as by current theory of evolution, accident is said to cause mutation and hence evolution. Since the quantum of uncertainty manifests in particles and atoms in ɪandom fashion, it may be thought of as accident, but this designation becomes inappropriate in later stages when higher organisms develop and invest the intrinsic randomness with a highly competent organization; for in this case, the intrinsic randomness seems more like "play" than accident, much as, in childhood, play results in mere accident, but in later life play can refer to the activity of a skilled athlete or a gifted virtuoso.

There is in all creation this transcendence of what is strictly rational or implied by its antecedents, and the word "play" comes a little closer than the word "accident" to describing the cause of new creations, whether they be those of the mathematician, the poet, the painter, or even the progress of evolution itself. The point is, we must free ourselves from the compulsion to view everything as having an antecedent or cause, or, in the name of science, to invert means and ends. It would not occur to us to say the creativity of Beethoven was due to his learning to play the piano; in fact, we would recognize that learning the mechanics of music was part of the means that Beethoven employed to transmit his creations. In a like manner, we should put first things first in interpreting nature, and think of the organisms as the means which the divine play uses to express its exuberance, rather than to presume that all of life can be explained as caused by laws.

Laws describe constraints; they do not create. We must posit a universe which does both, and in what follows we will endeavor to trace the interplay of these two primary forces—the creative play and the laws of necessity.

Since most science deals only in the latter, leaving evolution to be explained as accident, we can say that we are setting up a new paradigm or model for the universe, a paradigm which deals expressly with the interplay of freedom and constraint. But we cannot do this without careful attention to what science has found. This testimony is all the more valuable because, contrary to the former belief of science in the omnipotence of law, quantum physics has discovered that uncertainty or play is the most fundamental entity of all and is more basic than the fundamental particles it can create.

So we will follow the quantum of uncertainty as it falls into matter.

Chapters IV and V establish the broad outlines of the fall and subsequent ascent.

Chapters VI to IX fill in the details as confirmed in the evolution of atoms (VI), molecules (VII), plants (VIII), and animals (IX). Chapter X ventures into territory implied by the theory but not recognized by science, and Chapter XI examines animal instinct in the light of the principles implied in Chapter X.

Chapters XII to XV carry our thesis to its conclusion: the evolution of man and beyond.

IV | The four levels

In the concept of a fall followed by an ascent, process takes on a shape and becomes something we can describe.*

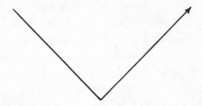

The description begins with the recognition of a purposiveness in process—a goal. To reach the goal, means must be found which to be effective must be predictable, must operate according to law.

Process and purposiveness

We thus by the single concept of purposiveness, account for a number of apparently different aspects: that process has direction, that it builds on itself, that it must use means, that means must be determinate.

We can now begin to feel a theory taking shape, and as I have noted in the Introduction, if we add that process has seven stages, we have a definite prescription to be filled; we have a set of instructions.

It was at this point that my investigations began in earnest. But before going further, I would like to plead the case for the experimenter

* We are now picking up the discussion from Chapter I, *The fall into determinism*.

with a theory, in order to correct what I believe to be an erroneous concept of how one proceeds with a theory. The usual assumption is that one adopts a theory and goes out to test it, and if it breaks down, one gets a new theory. This might indeed be the case, but, if so, only over a very long period. Most of the time it is just the other way about; when you believe you have a good theory you test it, run into trouble, then you *fix* the theory. It's very like a helicopter: the first time you rev it up, it has the shakes, so you study vibration problems until you find a solution. Or like raising a child: he fails his geometry—you don't get a new child,* you reinforce the weak point.

Actually, some research programs *are* operated on the alternative method: if a project is failing, drop the project. But little comes of this except the absorption of funds. Anything that counts is achieved only after much effort both creative and supportive, and I insist that this is the case with theories.

But as applied to cosmology, these observations may seem like generalizing without adequate evidence, for there have not been many overall cosmological theories to judge by. Current theory, with a great show of learning, has moved from naïve interpretations of the myth of Genesis to naïve interpretations of 19th-century science and has left it at that. There is no comprehensive cosmology based on quantum physics. Almost everyone, within the scientific disciplines, from behaviorists to physicists, is mentally conditioned to think in terms of classical science and fails to appreciate the possibilities offered by quantum physics for a theory which can transcend the limits of classical determinism.

In any case, we now have a theory, or a basis for a theory, and I would like to take the reader through the steps by which I forced out** of this theory the reasons for its "structure." (I put this word in quotes, for I've already explained the inadequacy of structure to describe the true development of process.) I use the word structure to denote the relations of symmetry and other properties of what I call the levels through which process descends and subsequently ascends, as we will demonstrate.

---

* Newton flunked his mathematics.
   ** This forcing was because I kept the same theory and made it talk to me.

Loss of freedom (an illustration)

In the descent, process loses its freedom in three downward steps. Let us imagine that you are trying to capture a wildcat that has climbed a tree. You lasso him with a rope and make the rope secure. The wildcat can still move about, but he can't get away. Then you lasso him again and make the second rope fast. The wildcat can still move, but whereas his movement was first confined to a sphere, the pull of the two ropes will constrain it to a circular orbit on a plane. (A circle is the locus of a point equidistant from two given points.) A third rope will complete the process and hold the wildcat in one position.

   Similarly, the step down from light to the level of nuclear particles constrains the particle to motion within a sphere (which is the orbit of uncertainty of the electron as described by Heisenberg); the second downward step confines the electron to movement in a circle around the nucleus of the atom; the third to a fixed position of the atom, as in a crystal.

Levels of descent

The descent is not continuous; it occurs in three steps. But why necessarily three? In the case of the wildcat, it is clearly because there are three dimensions of space. In the case of process, this three-dimensionality carries over into something more abstract, but still threefold. To demonstrate, I would like to present three ways of describing the entities of physics in four stages involving three downward steps. The individual levels will be described in more detail later in the book; here we are concerned only with a method for structuring this concept.

1. Division of initial unity

Described one way, the descent is, as we've mentioned (see pp. 18ff.), a *division* of the initial unity or wholeness in the quantum of action into Energy times Time, and Energy into Length times Force.

2. From homogeneity to heterogeneity

But there is another kind of division. This is the decline from homogeneity to heterogeneity of the entities themselves:

1. One kind of *photon*, which has unit spin* and no charge.
2. Two kinds of nuclear *particles*, which have half-spin and are charged positively or negatively.**
3. One-hundred-odd kinds of *atom*, with various chemical properties.
4. Countless kinds of *molecule*, with many kinds of properties: mechanical, electrical, chemical, physiological.

3. Change in degrees of certainty

The levels also represent the *degrees of certainty* which it is possible to have about the entities at the respective stages of process, and these degrees can be correlated to electron volts, a measure of energy.

Level I. The *photons* at level I are complete in their uncertainty: they are unpredictable. As we have noted, the observation of a photon annihilates it, so that there is nothing left to predict. The energy of a photon that can create a proton is about a billion electron volts, to create an electron about one-half million volts. All photons have total freedom.

Level II. The *nuclear particles*, electron and proton, created by photons, are the first occurrence of permanent mass and charge, the basic substance of the universe, as compared with the activity of the light that created them. But not all of this activity (or, more correctly, angular momentum) is condensed into mass. For reasons which are still unknown, $1/137$ of the angular momentum remains uncommitted, and free. (This $1/137$ is known as the fine structure constant.) It is this "freedom" which manifests itself in the uncertainty of position and momentum that characterizes the fundamental particles.

---

* Spin is an additional property of particles.

** I refer here to the permanent nuclear particles, proton and electron.

Level III. The *atom* entails a further reduction, not only in the sense that the charge of the contributing particles is neutralized, but in that the free energy which it radiates or absorbs is drastically reduced to about 10 electron volts (for hydrogen).

Level IV. In the *molecule*, it is the *bonds* that have energies, which cover a wide range. We will be interested in the energies of approximately 1/25 of an electron volt, which is that of the average molecule at room temperature. Why? Because according to our theory, it is at this energy level that life becomes possible.

This last energy level, it would appear, is the working base that process has to reach before it can start building up again. By this, we mean building the complex organic molecules such as proteins and DNA that are the basis of life, which requires a temperature between 0° and 45° centigrade.

Necessity for free will

The reader may perhaps be familiar with current ideas of how life arose from electrical discharges in the early atmosphere of methane. Such discharge has been experimentally tested and found to produce minute amounts of many of the amino acids necessary to life. This may indeed have been an important step in the creation of life, but in our view it could by no means have been sufficient. Our position is that life requires, in addition to materials and conditions, an *act of will* comparable to the purposiveness of the quantum of action. I am afraid this will seem unscientific, but I hope to show the contrary: that the hypothesis I am setting forth requires a minimum of assumptions. And it does not hide the problem under the rug, as do current interpretations of the Darwinian theory.

Neither of the preceding ways of distinguishing the levels lends itself in any clear-cut fashion to the higher kingdoms, plants, animals (man). (I am putting man in parentheses to indicate that he is not to be thought of as the sole representative of the seventh kingdom. In fact, part of the job is to arrive at a definition of the seventh kingdom, or at least a description that will distinguish it categorically from the others. It is not enough to make man a "naked ape" or "an ape with a club" or even an animal that "communicates through language" or "is capable of abstract thought." We have in the theory a tool that provides for more basic distinctions, and we should use it. This is what the levels provide. And it is perhaps not too early to point out that what is emerging is a scheme of the cosmos in which life, far from being "a green scum on a minor planet," as one scientist put it, is inherent in several *levels of organization that are intrinsic to cosmology.* The seventh kingdom, which includes man, is one of these levels of organization.)

So how can we show that the higher kingdoms are categorically distinct and occur *on separate levels?*

Symmetry of descent and ascent

There are several possibilities, but the one that is clearest, because it is visual, draws on the well-known, though neglected, *symmetries* of minerals, plants, and animals. This was first brought to my attention by Fritz Kunz, who discussed the subject in his article "On the Symmetry Principle."*

While D'Arcy Thompson in *Growth and Form*** devotes a large part of his thousand-page work to the subject of symmetry, he does not appear to notice the eloquently simple fact that the kingdoms may be distinguished by symmetry.

Molecules: level IV, stage four

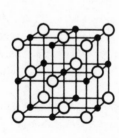
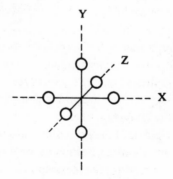

* Kunz, Fritz I. "On the Symmetry Principle." *Main Currents in Modern Thought*, vol. 22, no. 4 (March–April 1966).

** Thompson, D'Arcy W. *On Growth and Form.* London: Cambridge University Press, 1917, 1963.

Crystals, in the molecular kingdom, have what is known as *complete* symmetry: they comprise molecules in rows, columns, and layers monotonousiy and indefinitely repeated; *three* axes of symmetry and no freedom. By the term axis, I mean a direction of symmetry.

It seems beyond comprehension how a virus molecule, having a molecular weight in the millions, can be marshaled into the strict order of a crystal, yet this is true. All molecules can occur as crystals.

Plants: level III, stage five

Plants exhibit their one degree of freedom by growing vertically. The top of the plant differs from the roots, but right and left are similar, front and back are similar. This is known as *radial* or cylindrical symmetry; *two* axes of symmetry.

Animals: level II, stage six

Animals differ front and rear, and also top and bottom, but are similar right and left. This is *bilateral* symmetry; *one* axis of symmetry.

Note that the greatest symmetry (three axes) occurs with molecules, the least with animals, suggesting a *correlation of symmetry with constraint.* We have here a simple way to make a *quantitative distinction* among these three kingdoms by counting the axes of symmetry, or conversely the degrees of freedom. We may say of crystals that they have no freedom, while plants have one degree of freedom (their ability to grow) and animals two degrees of freedom (their ability to move about two-dimensionally on the surface of the earth). The flight of birds is also two-dimensional in the present sense, since birds steer vertically or horizontally.

But there is a position open for a kingdom with three degrees of freedom and no symmetry. (Kunz indicates a tendency toward asymmetry in the human face. There is also left- and right-handedness and the recently discovered fact that the two sides of the human brain have different functions. As already mentioned, however, the seventh kingdom is not to be thought of as limited to humans.)

The question now arises: can a similar quantitative distinction be applied to the left-hand side of the arc—to atoms, to nuclear particles, and to light?

Atoms: level III, stage three

While the image of electrons traveling around a central nucleus like planets around a sun has been supplemented by more abstruse models, the radial symmetry of the atom still holds, as is brought out by its magnetic properties.

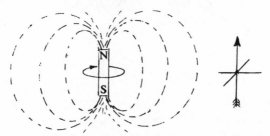

Diagram of a magnetic field

What gives the atom one degree of freedom (like the plant) is the fact that it can absorb or release energy without any prompting from outside. Its energy state is unpredictable (or free). The same can be said of plants, since growth correlates with energy storage by carbohydrates, etc.

Nuclear particles: level II, stage two

Let me first take up freedom. Recall Heisenberg's observation that we are uncertain of the position and momentum of the nuclear particle: hence it has two "degrees" of freedom. The product of the uncertainty of position and momentum is a unit of action and may not be less than the value h. The formula for this is $L \times ML/T = ML^2/T = h$. This situation is similar to that of an animal at large: we can know only an area (L^2) in which the animal (M) would be after a given time ($1/T$). The product cannot be less than a given value, $ML^2/T = h$.

As to the symmetry, this question cannot be answered with finality, but the experiments proposed by Lee and Yang, and completed by Mme. Wu, which discovered that chirality, or handedness, characterizes nuclear particle reactions, suggest this possibility, for only that which has bilateral symmetry can have handedness. (One could not have a right-handed circle or cone, but one can have a right-handed thread or spiral.)

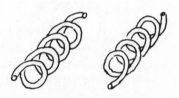

Light: level I, stage one

To carry out the scheme, we should show that light has no symmetry and complete freedom. I can't see how to establish its asymmetry.* Light is certainly the most completely free form of existence there is: a photon released at a certain point could be anywhere within a radius

---

* We might point out that symmetry requires measure, and light is before measure.

of 186,000 miles a second later. In addition, we can again point out that since observation annihilates the photon, it cannot be predicted.

As for the seventh kingdom, we will make lack of symmetry and complete freedom a *definition* of the kingdom. Since this is the highest form of existence, we cannot expect to define it anyway, and this, while a negative definition, is as good as we can expect.

We may put this all together in a chart:

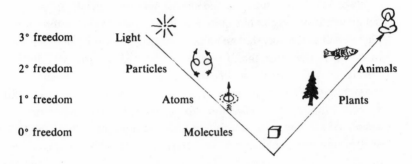

3° freedom	Light	
2° freedom	Particles	Animals
1° freedom	Atoms	Plants
0° freedom	Molecules	

Anticipating the character of the kingdoms

Now let us take advantage of this charting. Notice that there is a considerable economy of definition. The "degrees of freedom" describe the levels applied to kingdoms at the same level on both the right and the left side of the arc. And we can extend the correspondences further. Thus it would be expected of the atom that its position can be determined, as can that of plants. This is the case: the atom can be fixed in a crystal. The nuclear particle, unlike the atom, should have a fixed (internal) energy. This is the case: the energy fixed in the *mass* of the proton or electron is quite exact (known to the fifth decimal point). Correspondingly, the animal, once grown to maturity, retains a fixed weight in a way that is not true of the plant, which continues to grow.

A further point: animals and plants *interchange* position and energy in respect to which is free and which is constrained. Thus the animal is free to move but unable to create its own energy, while the plant is fixed in position but able to synthesize energy from sunlight.

Likewise, nuclear particles are free to move in the sense that their position cannot be determined. But their mass is fixed, while the atom, which can be fixed in a crystal, is able to absorb or radiate energy.

As we have noted, motion requires two degrees of freedom, while energy absorption requires one degree of freedom. Of course, it was the descent or "fall" into constraint followed by an ascent that was our basis for the arc. The inclusion of degrees of freedom adds precision to this conception by providing a quantitative measure.

We can now recognize that the kind of determinism which the behaviorist talks about has quite a different meaning from that of the physicist. Thus, to say that an animal is attracted by food or conditioned by drives in the sense that a weight is attracted by gravity is quite erroneous. The animal may be attracted to the water hole because of thirst, but it isn't going to move in accord with an exact "law" as would a freely falling weight, *subject to a force proportional to the inverse square of the distance*. A drive, such as hunger, does not wholly determine animal behavior as law determines the behavior of an inert object.

The reflexive universe

We have discussed similarities between the left and right sides of the arc. What is the *difference* between the two?

Notice again that the entities of exact science are on the left. Life, on the right, is not accounted for by science.* The higher kingdoms on the right have acquired an ability not present in those on the left—an ability we describe as *voluntary*, as distinct from *random*. Thus the movement of animals is voluntary, whereas that of nuclear particles is random; the storage and release of energy of plants is voluntary, whereas that of atoms is random.

This distinction, obvious to common sense, is difficult to maintain in the framework of current science because there is at present no formal expression for control. Such an expression can, however, be developed with the tools available from science.

* Biology, the "science" of life, is solely descriptive. It does not explain the life force, why things are alive, or why they move.

Formal expression for control: position and its three derivatives

Science depends on providing formulations for describing otherwise elusive concepts. Recall the paradox of Zeno which pictured a race in which Achilles was to overtake the tortoise, who had a head start. Achilles could never pass the tortoise since he would first have to get to where the tortoise was, by which time the tortoise would have moved on, and so on. Silly, but without a formal expression for velocity, which would permit a comparison of rates rather than of positions, the problem remained baffling.

Newton provided the necessary formal expression in the calculus, where he defined *velocity* as the rate of change of position with respect to time, and *acceleration* as the rate of change of velocity with respect to time.

Velocity is known as the *first derivative* (of position), acceleration as the *second derivative*. These two expressions laid the basis for the theory of gravitation.

While Newton mentioned a third derivative, he made no attempt to give it a physical meaning. What is it? Since each derivative is the rate of change of the quantity derived (i.e., velocity is the rate of change of position, acceleration the rate of change of velocity), we may conclude that the third derivative is the rate of change of acceleration.

Every automobile driver has direct experience with the third derivative, for in controlling the car by pushing the accelerator, applying the brake, or changing its direction with the steering wheel, he is changing its acceleration. This, in fact, is *control*. So we can say, just as acceleration is change of velocity, so control is change of acceleration and *is* the third derivative, and hence has status.

The neglect of the third derivative by classical physics is traceable to the fact that it cannot be used for prediction. We may, of course, as in a guided missile, lock the controls to home in on a target and hence render control determinate, and this is the special case covered by cybernetics. But in the general case, we must go a step further and recognize that control is "outside the system." It is indeterminate—the driver is free to steer the car where he wishes. This does not deny its existence as a factor in evolution.

We can diagramatically represent the derivatives by a circle on which position is shown at the right and its three derivatives in sequence clockwise.

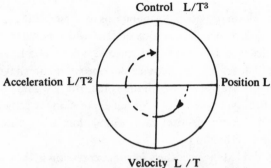

Such a circle is also representative of the cycle of action, and applies to any repeating cycle, such as the swing of a pendulum.

Here, however, we are interested in the fact that the representation implies that derivation returns to itself after four applications. Is this the case? Does the fourth derivative reduce to a position? Yes.

For example, when you're driving a car, your control of the car is governed by position, for that is what your destination is, a position in space. Or again, the control of a guided missile is directed by the position of the target. Therefore, *the fourth derivative is position.* In other words, if we divide by T four times, we are back at the start: $1/T^4 = 360° = 0°$. (Standing still, known as the identity operator in science.)

We propose to make *control* a criterion for the description of entities on the right-hand side of the arc (see p. 41). Our right to do this stems from the fact that control can be identified with the third derivative and is therefore equal in status with other derivatives (velocity and acceleration). Or, again, control is evident to observation: an automobile, a paramecium, a flying saucer can be observed to be under control or not under control. And control is evidence of life.*

---

* I have recently learned (January, 1971), thanks to an interesting volume by Marjorie Grene entitled *Approaches to a Philosophical Biology* (New York: Basic Books, 1968) of the ideas of a German biologist, Helmuth Plessner. Plessner's concepts stem from the recognition of the way in which an organism bounds itself, its "self-limitation," which Grene translates as "positionality." Not only is this quite similar to what I call "control," but it leads Plessner to postulate three *levels* of "positionality": vegetation, animals, and humans.

V | The turn

Out of the nettle, danger [determinism]
Pluck the flower safety [freedom]

   Apologies to Hotspur, in *Henry IV*

Requirements for the turn

The discovery of a third derivative is important, for it shows how we
may draw on the foundation of Newtonian mechanics for a very
different conclusion from the determinism it was thought to imply. The
third derivative, which we have called control, describes the control of
acceleration (or of force, if we add mass to acceleration). This is
action, a rebirth of the first cause at level I.*

   But our task is not over; in fact, we have done little more than to
advance our physics up to the sophistication of our instinctive
inheritance, for we continually have access to control, not only in
driving automobiles but in all kinds of motion: walking, talking, or
even just standing upright. What we must now show is not only that it is
possible to control machines, but that it is possible for a living organism
to control itself.

---

* First cause with the photon at level I is action ML^2/T. In this physical world,
first cause is the implementation of self-control. Control $= L/T^3$, force $= ML/T^2$.
Their product is ML^2/T^5. But $T^4 = 1$, hence $ML^2/T^5 = ML^2/T$, which is action,
not to be confused with the simpler control of force, or ML/T^3. To put this in
English: the physical world is the state of affairs in which the parts have been
separated out; it is populated with objects, forces, distance, time. But when we
know how to control force (when we multiply $L/T^3 \times ML/T^2$), we create the
"turn."

We must discover, now that we have control as an option, just what is required in order that control be possible. Given a machine that can be controlled, and given a hypothetical monad*—a ghost, if you like— how can the ghost control the machine?

Here is where the arc makes intelligent handling of the question possible, for the arc shows that it is at level IV that life begins and that it is at that point we should look for the control situation.

This is the realm of molecules and so, on the basis of the arc, we can assume that there is a point in molecular evolution when a molecule starts to build up energy and begins to move against entropy. The *polymers* are an example of just this phenomenon. They build themselves, and the process, unlike a crystal, can be *endothermic*: it can store energy and cool the environment.

Polymers (chain molecules) are not life in the usual sense, even though they grow and replicate themselves; but they show that a molecular phenomenon that reverses entropy can occur.

The beginning of life

This is the crucial step, for it implies that the monad, after its long precipitous descent from an energy level of a billion electron volts to the fraction of an electron volt involved in these molecular bonds, has at last taken the wheel and begun to climb out of the abyss.

I am not particularly disturbed by invoking a sort of molecular intelligence here. We must understand that our cosmology should provide the necessary equipment for this crucial step, even if it doesn't provide or explain the act of will which brings it about. This bottom of the arc is the prodigal son situation. The return does not come about *because* of the law; it must be voluntary, and has to wait until this point because not until now is there a basis on which to build a return.

Molecular bonds and temperature

We have already mentioned molecular bonds at level IV (p. 36). These bonds are, as it were, the "last stand" of the quantum of action which

---

* This is my first use of the term "monad." Up to this point it was the quantum of action, but now it acquires status as the vital spark in life.

here, at the molecular level, acts to hold the parts of a molecule together. We may think of this bond as an energy, because it requires energy to pull apart that which the bond holds together. Such molecular bonds, once formed, do not of themselves come apart. However, they may be separated if the molecule is struck hard enough by another molecule. How hard must it be struck? Enough so that the energy of impact is equal to the energy of the bond.

It is at this point that the temperature of the environment comes into play. The environment consists of other molecules batting themselves about in a random fashion, and the temperature of this environment is the rapidity of this motion, which imparts to each molecule an average kinetic energy which is greater as the temperature is greater. When the temperature is high enough so that the energy of the average molecule approaches that of the bond, the molecules begin to split up. It is this fact that is made use of when metallic ores such as iron oxide are heated in a furnace to drive off the oxygen and release the pure iron. Or again, where crude petroleum is heated and the heavy-oil molecules are broken down into smaller gasoline molecules. Many chemical operations make use of this principle that chemical bonds are broken by the use of heat. The cooking of food is a familiar example.

Now we know that life is peculiarly sensitive to heat. Cooled to the temperature of a household refrigerator, most life forms cease their processes—including the activity of decay bacteria; the food is preserved. Heated to a temperature somewhat over 100° Fahrenheit, most life ceases. Milk is pasteurized at 140°, which kills the disease bacteria. This temperature range for active life—from approximately 23° to 221° Fahrenheit—is a relatively narrow one. For warm-blooded animals the range (excluding hibernation) is still smaller, from 94° to 113° Fahrenheit, a tiny span if one considers the range between absolute zero (−459.4° F) and the temperature at the surface of the sun (10,800° F).

Life can exist only within this narrow range. Why? The answer must be that the chemical bonds on which life depends cannot be formed at temperatures below the lower limit and cannot be maintained above the upper limit. Life requires this temperature to make the energy transactions that maintain it.

We also know that the most basic characteristic of life is that it can reverse entropy, that is, it can store order in contrast to the tendency of nonliving matter to reduce order (or increase entropy; i.e., stones tend to roll downhill, not uphill). Thus vegetation takes energy from sunlight to store it in starch and sugar, and animals consume vegetation or other organic material to rebuild it into their own bodies. Neither one creates the energy they store; they extract it from their environment.

This is what life does. We can see that it occurs, and we can even analyze the steps by which it occurs, much as we can follow the steps by which a carpenter erects a building out of the material he obtains at the lumber yard. But this does not explain the whole process either as life does it or as the carpenter does it. To explain the carpenter, we can speak of "purposive action," but the dictates of 19th-century mechanistic science (determinism) forbid our speaking of such purposive action at the molecular level. (Indeed, they forbid our speaking of it at any level. But at the human level—that of the carpenter —we have the direct testimony of experience which overrides the theoretical assumptions of determinism.)

Quantum mechanics brings forward new evidence to show that the prohibition of mechanistic science is invalid for individual particles. Electrons and atoms bounce about in unpredictable and random fashion; their behavior cannot be distinguished from that of free entities.

How does this new ruling apply to molecules? Here is where the temperature factor is important. For there is a temperature range in which the molecular bond behaves in a free or random fashion. It can break apart or not break apart, depending on whether it is *in phase* or *out of phase* with the incident energy that is bombarding it. This point is self-evident, but the interpretation which makes it the entry point of free will (see p. 46) is my own.

The molecule is in this respect like a wrestler who can react either with or against his opponent and by correct timing defeat a heavier man.*

The choice of timing is the only freedom left to the molecule at substage four, and it affords the opportunity for the monad to reverse

---

* This is the kind of action in which Maxwell's demon engaged. But note that the demon, like a wrestler, *interacts directly*; he is not required to observe and compute as is his more intellectual counterpart—the Heisenberg observer—and thus does not require energy. This disposes of the current claims (based on energy requirements) that the demon has been exorcised.

its fall into determinism and, by collecting energy, to create the organization (substage five) which will begin the ascent back to freedom, this time with greater scope.

Timing is the ingredient that marks success in any field, whether it is the wrestler who overcomes a more powerful opponent, the success of a business venture, the performance of a musical composition. It is not energy; it is not force, nor even the control of force; it is the correctly timed control of force, and it has to be learned.

The fact that it has to be learned is important, for it explains the descent (the fall); it explains why process goes so far out of its way to achieve its ultimate goal, and it explains why the world is as it is.

The "turn," then, is the origin of life and it depends on timing, the hidden freedom of level IV. The turn is the most important point of the whole arc, and while we cannot completely explain it, it is essential that we make room for it in our theory.

We should now implement the above heuristic description with a more formal explanation. This is possible now thanks to Eddington's profound insights reconciling quantum theory and relativity. Since this involves concepts difficult even for the physicists, the general reader may prefer to move on to the next chapter (Atoms). I would urge him, however, to note the section entitled "Comparison with classical physics," which accents important points by comparing the present theory with the older view of a universe determined by laws but unable to account for life.

Explanation of the phase dimension

Perhaps the greatest contribution of the theory of relativity is that it adds a new dimension to the known physical universe. This dimension is not time, despite the impression we may have received from popular accounts of time as the fourth dimension. Rather, it is what Eddington calls *phase*.* Eddington goes on to show that, beginning with the

* This is not phase in the sense of Gibbs. It is rather phase in the sense of a lead or lag, as in the case of alternating current, in which voltage and current may be out of phase.

ordinary physical universe represented as a sphere, when we include phase we widen the universe in another dimension perpendicular to space–time.

An additional factor of 3/4 must be introduced to stabilize the scale. To get the higher dimensional universe (the Einstein hypersphere; see Appendix II), we multiply the volume of the three-dimensional sphere by these two factors, 2π and 3/4:

$$2\pi \times 3/4 \times (4/3)\pi R^3 = 2\pi^2 R^3$$

The phase dimension, an uncertainty of 2π, describes the maximum possible range of angle, 360°. It is a circle. And I would add it is also a cycle, the cycle of action, and stands in place of time. We might have expected the formula for the universe to express the time dimension as well as the three spatial dimensions, but instead it describes time in much the way that we do when we speak of a person's lifetime, not in years but in terms of birth, youth, maturity, middle age, etc.

This extra phase dimension is the observer's uncertainty as to which direction a thing will move. As Eddington points out in *Fundamental Theory* (see Appendix III), it applies to each particle or small system. Here in what is surely one of the most remarkable passages in scientific literature, he opens his Chapter III by dropping the curved space of molar relativity and replacing it by the uncertainty, that is, by the curvature of 2π.

Since 2π is both a *curvature* of space in the hypersphere of relativity, and the *uncertainty* of direction in quantum theory, Eddington, by recognizing their equivalence, is able to reconcile quantum theory and relativity. Einstein had attempted this reconciliation in his quest for "a unified field theory," but was looking in the wrong place; he overlooked the fact that it was not the field but the uncertainty that could accomplish this task. In fact, Einstein in his famous dictum, "God does not play dice with the universe," banned the very factor, uncertainty, which could effect the reconciliation.

Important as this point is—and I'm afraid it is sadly neglected in high quarters—our task lies in going still further, in the recognition of the significance of this 2π, whether we call it uncertainty or curvature. Eddington comments on the fact that "now that each particle or small system has its own scale variate, a new field of phenomena is opened

to theoretical investigation, which is suppressed in the molar treatment."
This "new field of investigation" could well be the present study, the
investigation of process, or the implications which follow from the
postulate that the quantum of action (which includes the 2π
uncertainty) is the active core of a particle or an organization of
particles, the ghost in the machine.

It is our hypothesis that this range of angle, which to the observer is
uncertainty, is a *range of choice*. We may think of this choice as either
a choice of direction or a choice of *timing*.

But we should not neglect the other factor of 3/4. This factor is what
Eddington calls "stabilization of scale." From our point of view, this
stabilization of scale is "self-limitation" or *control*, the same control we
referred to on page 44, which arises in the cycle of action at the
point 3/4 of the way around the circle or cycle of action.

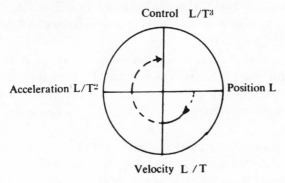

Control L/T^3

Acceleration L/T^2 — Position L

Velocity L / T

To assist in the understanding of this crucial point—that control is
3/4 of a circle—we can invoke what is known in psychology as the
learning cycle. The learning cycle refers to the trial-and-error process
by which a child learns to avoid a hot stove, a rat learns to tread a maze,
or even a flatworm to avoid an electric light.

This cycle begins at (1), when random action (equivalent to
acceleration) encounters some painful experience, say a hot stove; it
then reacts by withdrawal at (2); and at (3) associates the pain with
the stove (conscious reaction). After this sequence, it can, at (4),
consciously* avoid the hot stove. This is *control*, but it is achieved by

---

* If one objects to the word "consciously," we can say, with the behaviorist, that
the pain is associated with the hot stove.

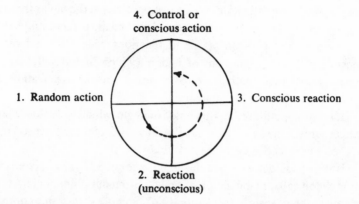

a different route from that of the derivatives: it moves counterclockwise from (1), whereas the derivatives proceeded clockwise from (3). The derivatives are the basis for sophisticated or informed action, whereas the learning cycle is the reverse of this, the process by which control is learned in the first place.

While this need not be consciousness as we know it—its duration is still but a trillionth of a second (ours, it would appear, is about 1/10 of a second or less)—yet it is the consciousness appropriate to a molecular bond, meaning that by control of phase or timing, it can lock or unlock such bonds.

This range of choice is indeterminate from the point of view of the observer and, by the same token, is freedom for the "entity," i.e., the bond. In other words, the quantum of action is now able either to absorb or to release energy and hence is able to break the inevitability of the second law of thermodynamics (increase of entropy).

We must be careful, however, to distinguish this from the *reversal* of entropy, which can be achieved by purely mechanical devices. A self-winding wristwatch is an example. Here the random movements of the wearer's arm serve to induce rotation which winds a spring. The opposite rotation, which would unwind the spring, is prevented by a ratchet. Such a device could also operate on a molecular scale, and may serve as another adjunct to life, but it does not explain the act of will to which we referred on page 36 and again on page 46. To account for the essential freedom of will, and the option of life to store and release energy, we require the free phase dimension. The free phase dimension, then, is that which makes it possible for the quantum of action to control *timing*.

Comparison with classical physics

Let us now consider the quite different view of the universe which the theory of process invokes, as compared with that of classical physics.

Classical physics	*Theory of process*
The universe is based on law.	The universe is based on freedom.
	Laws, constraints on freedom, are secondary.
It deals with molar objects (statistical aggregates).	It deals with individuals (quanta of action, single atoms, molecules, and organisms).*
Objects are fundamental.	Action is fundamental.

In the theory of process, laws remain important, but they do not have the omnipotence which classical physics assumes. Process, in fact, emphasizes the importance of nonlaw, or uncertainty, because of its positive contribution, especially to the emergence of life in higher kingdoms. (It is in these higher kingdoms that the change in degrees of certainty described in Chapter IV makes its major contribution.)

Additional information

Different kinds of law at each level
A further aspect of the theory, which can usefully be mentioned in conclusion of this chapter, is that the several levels exhibit different kinds of law.

It is interesting to note that we may characterize the four levels by the different degrees to which they obey the law. This shows up in a simple mathematical manner, in the degree of the relevant equations, whether linear or quadratic.

* The theory of process recognizes that an organism can be under the control of the quantum of action, as can a molecule. If this were not so, how could an organism grow from a single cell?

Level I. Here, if we are careful to realize what is going on, we find that, except for hierarchy, there are almost no "laws." The speed of light, as we said earlier, is not a restriction on light, but rather a boundary condition on matter; not a boundary to space, but a boundary to the combination of time, i.e., velocity. Law of distribution breaks down: $a(b + c) \neq ab + ac$. Equations here reduce to stating a constant. Action $= h$. Velocity of light $= c$.

Level II. The realm of linear algebra. Laws of conservation (mass, charge, spin, etc.).

Here there is one degree of constraint. The location of an event in time is an example. Laws here must in consequence be linear, laws of numerical summation.

$$A = B \text{ or } A = -B$$

In these expressions, A is either plus B or minus B—totally different. If we square both expressions:

$$A^2 = B^2 \text{ or } A^2 = B^2$$

they become the same. Thus the quadratic expression (the squared one) *conceals* the plus–minus-ness.

Dirac's device of expressing equations (formerly quadratic) in linear form uncovered the plus–minus-ness and enabled him to predict the positron. To anticipate later developments in this book, I may add that Dirac's contribution predicted a different *kind* of reality from that of the formulations of relativity; it has to do with substance rather than form (see p. xviii).

Level III. The realm of geometric and quadratic relation. Laws of motion, Coulomb's law, gravitation, etc.

Here we have two constraints, that is, two dimensions in which measures can be simultaneously taken, which is the condition for location in space, for description of shape. Thus measurement of space or the law of intervals (space plus time):

$$a^2 = b^2 + c^2, \text{ or } ds^2 = dx^2 + dy^2 + dz^2 - dt^2$$

But also Coulomb's law: the attraction varies as the inverse square of the distance. This world is that of concepts. Atoms (level III) are conceptual entities.*

Level IV. The physical universe. Boundary conditions.

Here is the regular physical universe which combines laws of location with

---

* A concept recently advanced by Walter M. Elsasser, *Atom and Organism: A New Approach to Theoretical Biology*, Princeton: Princeton University Press, 1966.

laws of time. Two bodies cannot occupy the same position at the same time (or there will be collision).

Even more simply, we can say of this world of level IV that it deals in particulars. "The pen I'm writing with" (*now* is understood). "Pen" is conceptual, as is "writing," but the use of the present tense locates the action in time and makes it particular. It also indicates a physical object. This level adds what are called boundary conditions to the laws of level III (laws of form).

VI | Atoms

The third kingdom, that of atoms, introduces the possibility of identification, of location in space, of clear definition. Here we have the atom: it is carbon, or oxygen, or sodium; we can locate it, tag it, and know its properties. In fact, thanks to the same quantum theory that denies complete knowledge of the proton and electron in the second kingdom, we have a complete explanation for the atom in the third kingdom. Not only does every atom (or element) have properties that can be completely explained by reference to the number of proton–electron pairs, called the atomic number, but the spectral lines radiated by the atom, which represent energy differences between orbits of electrons, can also be precisely accounted for.* The superbly rational properties of atoms are a characteristic we believe to be categorical, for, as we will later endeavor to show, the level of this kingdom (level III) is that at which concept formation occurs; more simply, it is *form* itself.

There are only about one hundred kinds of atom. I say "about" because those heavier than radium (atomic number 92) do not appear in nature, and, when created artificially, have lifetimes which become shorter and shorter as the weight increases, until it is only a fraction of a second. There is therefore an upper limit—hence a finite number of kinds of atom.

---

* The frequency of the spectral lines is predictable, but *when* the atom will radiate is unpredictable. This is the one degree of freedom to which we referred in our discussion of atoms in Chapter IV.

Rules for constructing atoms

We may note certain rules:

1. Atoms are constructed out of protons and electrons in equal numbers, or pairs—the heavy protons constituting the atom's nucleus, and the lighter electrons moving around it.
2. The number of proton–electron pairs determines the kind of atom. There is a different number for each atom (the atomic number) and a different atom for each number.
3. On account of the repulsion between positive protons in the nucleus, nuclei which contain more than one proton require an additional binding factor, or glue. This is provided by the addition of *neutrons*, which consist of an electron and a proton united together. There must be at least as many neutrons as protons for the nucleus to be stable.

We can now construct a few atoms:

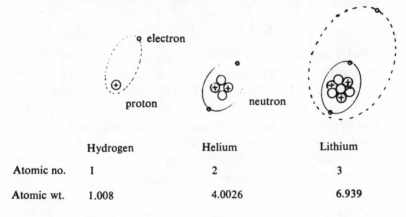

	Hydrogen	Helium	Lithium
Atomic no.	1	2	3
Atomic wt.	1.008	4.0026	6.939

We now encounter another important rule of atom-building. The added electrons fill up shells or rings, and the rule is that the first shell has two electrons which, when filled, is followed by another shell of two, then by a shell of six, then by another of six, then by a shell of ten, then another ten, then a shell of fourteen.

Angular momentum in the atom

It is the interpretation of these shells that has prompted the study of the atom. Because of them, the atom radiates and absorbs light of certain exact frequencies which show as lines in the spectrum (bright lines for radiation and dark lines for absorption). It was through deciphering these lines that the atom was understood.

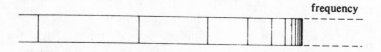

Their position, which could be determined empirically with great accuracy, was not accounted for theoretically until Balmer, who was not a physicist but a numerologist, in 1885 discovered the formula for the lines of hydrogen which gave their exact value and even predicted other sets of lines:

$$\text{Frequency} = 1/m^2 - 1/n^2 \qquad m,n = 1,2,3, \text{etc.}$$

But it was not until 1913 that Bohr, applying the concept of Planck that radiation occurs in quanta, gave an explanation of the behavior predicted by the formula (see pp. 14–16). Bohr assumed that the electron, like a vibrating violin string, can be in one of a number of discrete states, related by whole numbers, much as Pythagoras had found for the sound frequencies of a vibrating string. Bohr noted that when the electron jumps from one of these states to another, a quantum of energy in the form of a photon is emitted.

Then it was found that Bohr's explanation did not account for atoms with many electrons. Other explanations were advanced, and the quantum of action of Planck was found to be even more significant than had been suspected.

But the post-Bohr atom is not clearly explained in the textbooks. This is partly because so much was going on all at once. The de Broglie and Schroedinger wave equations and the discovery of the third and fourth quantum numbers (in addition to angular momentum and energy) overshadow what is important for the present theory: that in order for an electron to change orbits, the *shape* of the orbit has to change, and *shape is due to angular momentum, not to energy.* (See definition of angular momentum in the *Additional Information* section at the end of Chapter II.)

But why is shape of the orbit so important? Let us consider an atom in what is called the ground state. Here, the electron has no angular momentum. It is attracted to the nucleus like a moth to the flame, but its headlong dash toward the nucleus gives it such velocity that it is tossed back, like a comet passing the sun, as fast as it came, eventually to fall again, over and over. It cannot escape from the nucleus unless it receives a unit of angular momentum to give it circular motion and the option of choice instead of the in-and-out motion.

Ground state—no angular momentum   Excited state

If we liken the electron to an astronaut in a spaceship, we can realize that this motion around the nucleus, like an orbit around the earth, represents a kind of power, a control of the situation that the same astronaut in free fall toward the earth would not have.

We can illustrate the situation in yet another way. Suppose a man raises cattle in Texas and wishes to transport them to Chicago for market. He can drive live cattle to market and sell them, and he's paid for the meat, but not for the aliveness, though it is the aliveness that got the cattle to Chicago. So it is with angular momentum, which can convey energy, but is not energy. It is the wholeness, the aliveness, which disappears when the creature is no longer whole. In the atom, it is orbit shape. Thus I am stressing that the aliveness (angular momentum or the whole) is more important than the physical energy (the part).

So we must not draw the conclusion that angular momentum is a mere "epiphenomenon," in the way that aliveness is regarded simply as the by-product of more basic causes. Here at atomic and subatomic levels we find it has an absolute value—$h*$—and it invariably possesses energy. While this energy is very small, it is not necessarily so; in cosmic rays it is quite large. But, in any case, this energy is the right amount for what it has to do; it is commensurate with the atoms and electrons whose state it changes.

In thus calling attention to angular momentum, I want not only to stress that angular momentum shows up as having a primary role in the world of the atom. I also want to indicate how the challenges of quantum physics have led science, quite against its first convictions, to new concepts that are different from its earlier rational materialistic ones. Here we have a direct revelation that energy, unless carried by angular momentum, cannot be transported from one atom to another.

The reason I say revelation is that, like revealed religion, it comes from a source that is not in the ordinary run of affairs, and like revealed religion, it has to be interpreted.

Above all, I stress that we must make the most of this clue, deciphered with such difficulty from the cryptic utterances of the atom, and apply it. For if it is true in the minute world of atoms, it could well be true in the larger world of people. But we need to inquire further into the nature of the atom.

The periodic table

This brings us to the periodic table of the elements, whose discovery is credited to Mendeleev (1869). By its help, Mendeleev was able to predict the properties of elements that had not then been discovered. He called it "periodic" because it showed the periodic recurrence of similar chemical properties. Even before this, Newlands, in 1863, suggested a table of seven rows and seven columns, which was an oversimplification, but neither Mendeleev nor Newlands was correct from the fourth row on, because it has been found that the 8-electron shells that were first proposed are actually composed of a 2- and a 6-electron subshell, to which are added, in the fourth and fifth rows, a 10-electron subshell, and in the sixth and seventh rows, a 14-electron subshell. The 10- and

* See discussion of Planck's constant, pp. 14–16.

14-shells are closest to the nucleus, buried under the 2- and 6-shells, and hence do not influence chemical properties. That is one reason why they were not taken into account by Mendeleev and Newlands.

For simplicity, let's look at the first three rows of the periodic table:

1 Hydrogen						
3 Lithium	4 Beryllium	5 Boron	6 Carbon	7 Nitrogen	8 Oxygen	9 Flourine
11 Sodium	12 Magnesium	13 Aluminum	14 Silicon	15 Phosphorous	16 Sulphur	17 Chlorine

This is also the first three rows of Newland's table. With few exceptions, it includes the elements most important to the chemistry of life. In this table we can see the combining ratios, i.e., how to combine atoms in such proportions so there will be a total of eight electrons in a stable molecule.

Example

1st column atoms with 7th column in the ratio 1:1   NaCl (sodium chloride)

"   "   "   " 6th   "   " "   " 2:1 Na_2O (sodium oxide)

2nd   "   "   " 7th   "   " "   " 1:2 $MgCl_2$ (magnesium chloride)

"   "   "   " 6th   "   " "   " 1:1   MgO (magnesium oxide)

Carbon, in the fourth column, combines with all the rest, including itself (diamond). These combining ratios hold good for elements of the same columns in all rows.

The combining ratios, however, deal only with the columns, and do not tell us about the rows. It is the rows that are the major divisions, and hence are the guide to the process of atomic evolution. We must therefore give consideration to the extra shells introduced in the fourth to seventh rows. By separating the added rows into blocks, it is possible to get a clearer picture of what's going on:

H 1	A. 2-subshells												D. 6-subshells					He 2
Li 3	Be 4											B 5	C 6	N 7	O 8	F 9	Ne 10
Na 11	Mg 12			B. 10-subshells								Al 13	Si 14	P 15	S 16	Cl 17	Ar 18
K 19	Ca 20	Sc 21	Ti 22	V 23	Cr 24	Mn 25	Fe 26	Co 27	Ni 28	Cu 29	Zn 30	Ga 31	Ge 32	As 33	Se 34	Br 35	Kr 36
Rb 37	Sr 38	Y 39	Zr 40	Nb 41	Mo 42	Tc 43	Ru 44	Rh 45	Pd 46	Ag 47	Cd 48	In 49	Sn 50	Sb 51	Te 52	I 53	Xe 54
Cs 55	Ba 56	La and 57 58-71	Hf 72	Ta 73	W 74	Re 75	Os 76	Ir 77	Pt 78	Au 79	Hg 80	Tl 81	Pb 82	Bi 83	Po 84	At 85	Rn 86
Fr 87	Ra 88	Ac and 89 90-102															

C. 14-subshells

Ce 58	Pr 59	Nd 60	Pm 61	Sm 62	Eu 63	Gd 64	Tb 65	Dy 66	Ho 67	Er 68	Tm 69	Yb 70	Lu 71
Th 90	Pa 91	U 92	Np 93	Pu 94	Am 95	Cm 96	Bk 97	Cf 98	Es 99	Fm 100	Md 101	No 102

Counting helium as containing the first 2-shell, there are seven 2-shells, five 6-shells, three 10-shells, and one (complete) 14-shell, occurring in nature. (The second 14-shell comprises artificial elements with short lifetimes.)

Build-up of shell structure

We may represent these shells pictorially by taking a typical element from each row. We show below the elements from the second column:

Helium Beryllium Magnesium Calcium Strontium Barium Radium

Note the manner in which the sequence builds up progressively:

1. First we have a 2
2. Then another 2
3. Then we have a 6 plus a 2
4. Then another 6 plus a 2
5. Then we have a 10 plus a 6 and a 2
6. Then another 10 plus a 6 and a 2
7. Finally a 14 plus a 10 and a 6 and a 2

The periodic table is trying to tell us something about process. Concerning the atomic kingdom, it says:

Process is cumulative, each stage including what went before.
Seven stages of process follow an a-a-b (1-1-2) scheme, each stage
repeating before a new development occurs.

Properties of the rows

To discover more, we must draw on the properties which distinguish elements in different rows. These properties are not chemical, at least in the sense that the columns determine chemical properties. But they are quite interesting and suggestive for further study. For this, we need the full-dress version of the periodic table, so that we have complete rows.

Periodic chart of the elements

Inert Gases

H 1																	He 2
Li 3	Be 4											B 5	C 6	N 7	O 8	F 9	Ne 10
Na 11	Mg 12											Al 13	Si 14	P 15	S 16	Cl 17	Ar 18
K 19	Ca 20	Sc 21	Ti 22	V 23	Cr 24	Mn 25	Fe 26	Co 27	Ni 28	Cu 29	Zn 30	Ga 31	Ge 32	As 33	Se 34	Br 35	Kr 36
Rb 37	Sr 38	Y 39	Zr 40	Nb 41	Mo 42	Tc 43	Ru 44	Rh 45	Pd 46	Ag 47	Cd 48	In 49	Sn 50	Sb 51	Te 52	I 53	Xe 54
Cs 55	Ba 56	La and 57 58-71	Hf 72	Ta 73	W 74	Re 75	Os 76	Ir 77	Pt 78	Au 79	Hg 80	Tl 81	Pb 82	Bi 83	Po 84	At 85	Rn 86
Fr 87	Ra 88	Ac and 89 90-102															

Lanthanum Series

Ce 58	Pr 59	Nd 60	Pm 61	Sm 62	Eu 63	Gd 64	Tb 65	Dy 66	Ho 67	Er 68	Tm 69	Yb 70	Lu 71

Actinium Series

Th 90	Pa 91	U 92	Np 93	Pu 94	Am 95	Cm 96	Bk 97	Cf 98	Es 99	Fm 100	Md 101	No 102

I will now describe the characteristics of the rows of the periodic table in a manner which will later show their correspondence to the seven kingdoms. I.e., the words here in italics characterize the powers of the seven grand stages or kingdoms.

Setting the noble gases (helium, neon, argon, etc.) aside (since they do not form compounds), we note that the first row contains hydrogen. Apart from its predominance as the most common element in the stars, hydrogen atoms are more than twice as abundant as any other element in the human body, and thus hydrogen is the most *basic* element, like the *light* that started it all.

The second row contains carbon, oxygen, and nitrogen, which, with hydrogen, are the elements which make up the carbohydrates, fats, and the bulk of the proteins. In other words, the second-row elements make up the *substance* of the body. These molecules are the building material and the fuel. They have no fixed character, no identity.

The third-row elements contrast sharply with this. They are elements which make molecules that have a special purpose and retain their *identity*. We can discover this by considering each in turn:

Sodium. Remains as sodium chloride, which functions to maintain sodium–potassium balance.

Magnesium. Outstanding in its special significance as the core atom of chlorophyll molecules.

Aluminum. Not present in the body.

Silicon. Important in shells of diatoms, calamites.

Phosphorus. Makes the DNA double helix; each helix is a row of phosphorus atoms (see p. 80). Also essential to ATP (adenosine triphosphate, a compound which transfers energy and is very important to metabolism).

Sulphur. Important as the "hook" that joins protein chains, and thus makes possible the elaborate molecular structure of the proteins.

Chlorine. Linked with sodium–potassium balance and other special functions.

The fourth-row elements also have special functions, but here it would appear that the *combining* motif is emphasized. The combining power of iron with the other chemically similar metals in the 10-shell group is important in alloys of steel, which are obtained by addition to iron of chromium, vanadium, cobalt, nickel, etc. Similarly, zinc and copper make the important alloy bronze; nickel and chromium are used for resistive elements in heaters, etc. For the most part, however, we have to turn to the chemistry of life to discover how the latent potentials of the different rows manifest. For example, in the hemoglobin of the blood, iron combines with oxygen at the lungs, which is given up at the muscles.

Of the fifth-row elements, the only one I've found that is essential to the human body is iodine. It so happens that iodine is essential to the thyroid gland, which regulates *growth*.

Of the sixth-row elements, I can only offer that gold is used in the treatment of arthritis, and hence pertains to *mobility*; but I feel that this example will not be convincing to the reader.

When we come to the seventh row, chemistry is transcended by radioactive tracer elements. Perhaps man, with the trail of unnatural pollutants he leaves behind, has some resemblance to these seventh-row elements, all of which are radioactive, and destructive to their environment.

Arc-like descent of atoms

We would expect to find some sort of descent in atoms, followed by an ascent. How is this manifested? The descent appears in the curve that represents binding energy in the nucleus of atoms.

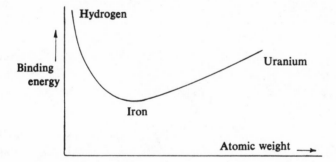

While this arc seems rather mild in its gentle contour, it packs a lot of wallop, for the difference between uranium and iron provides the energy release in the atom bomb (fission), and the difference between hydrogen and iron provides that of the hydrogen bomb (fusion). Also, iron is displaced to the left because there are more elements in the rows after iron (on the right). These factors adjusted, we would have a diagram that resembles the arc, and even suggests the evolution of the heavier elements from iron, still but partially understood, but thought to have occurred before the present sun existed.

To summarize: we have emphasized the degree to which the atomic kingdom can be rationally explained; how the variety of atoms is due to the number of proton–electron pairs constituting the respective atoms; and how the periodicity of this number, which creates the rows of the periodic table, divides the one-hundred-odd atoms into seven substages exhibiting, to some extent at least, the properties of the grand stages of process.

VII | The molecular kingdom

Divisions of chemicals

Whereas the atom *organizes* proton–electron pairs, the molecule *combines* atoms. The distinction is helpful and will be clarified in what follows.

There are only about one hundred kinds of atoms; there are countless kinds of molecules. Atoms as atoms exist only singly; when they combine, they form molecules. When molecules combine, the result is another molecule. If we make atoms analogous to the letters of the alphabet, then molecules, being combinations of atoms, are analogous to words. Atoms, like letters, are limited in number; molecules, like words, may be constructed endlessly.

Having found seven kinds of shell in atoms—a distinction on which the seven rows of the periodic table are based—*can we find a sevenfold division of the molecular kingdom?*

A number of years ago I asked this question of an outstanding chemist, Dr. Charles Price.* He supplied me with the following breakdown of the molecular kingdom into seven substages:

1. *Monatomic:* metals.
2. *Ionic compounds:* salts, most acids and bases.
3. *Nonfunctional compounds:* the paraffin series.
4. *Functional compounds:* compounds of compounds.
5. *Nonfunctional polymers:* chains 100,000 units long.
6. *Functional polymers:* proteins, which are chains with side chains.
7. *DNA and virus:* double helix, self-replicating.

In retrospect, it is surprising how pertinent Dr. Price's breakdown has been, for I have not had to change it despite new ideas, and

* Personal communication.

moreover I've learned from it ideas applicable to other areas of process.

I have found that this kingdom is very helpful to the study of process because it is more completely understood by science than other kingdoms.

These seven broad divisions of chemicals, like the rows of the periodic table, also have characteristics which illustrate those of the seven kingdoms. As we pass from salts (with two atoms) to DNA, with hundreds of millions of atoms organized into a single molecule, we find evidence for an extraordinary evolution. In what follows, I will summarize the basic plot of (this chemical) evolution.

First substage—metals

The essence of the molecule is not the constituent atoms, but the bond which holds the atoms together. For the simplest molecules, metals, which are monatomic, the metallic bond leaves the electron free to course through the body of the metal so that it can conduct electricity, as in the case of the telephone or power line. This conductivity of metals is appropriate to the first substage, for it resembles and includes the propagation of electromagnetic radiation at the first stage. (The metal wire is a wave guide for the signal.)

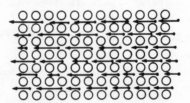

Metals: metallic bond
Electrons drift freely
through the metal

One might wonder why metals, in which the molecules are monatomic, are "molecular" at all; why are they not just atoms? Here we must recognize that the combination of atoms, of which this is the simplest type, results in something other than atoms. It produces metals which have properties that do not exist in single atoms. For example, density, malleability, melting point, conductivity, strength, etc., none of these "molar" (i.e., molecular) properties exists in single atoms.

Second substage—salts

Next in order of complexity are compounds of the simplest sort, such as NaCl (salt), which consists of two kinds of atoms, a positive one (sodium) and a negative one (chlorine), held together by their opposite charges. These positive and negative atoms resemble the second stage, protons and electrons with opposite charges. Salts are soluble in water and form readily into crystals in which atoms are so closely surrounded by their neighbors that it is impossible to say which atom goes with which to make a molecule. The order is such that each sodium is surrounded by six chlorines and each chlorine is surrounded by six sodiums. Thus we can say that this molecule of the second substage, having no identity, is bound to its neighbor in a collectivity.

Let us note the appropriateness of the term "binding" for the second substage. Here one atom gives up one of its electrons to the other. The former acquires a positive charge and the latter a negative charge. In this charged state the atoms are known as ions. The bond in this case is *ionic*; it comes about because of the attraction which these oppositely charged ions have for one another. But this attraction is communal—each positive atom in the crystal is surrounded by negative neighbors, and each negative by positive neighbors, hence *binding* the whole into a crystal. There is thus no identity to these molecules.

Second-substage molecule: salt (NaCl)

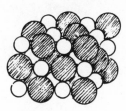

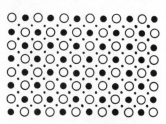

Salts: ionic bond Electrons confined to a neighborhood

One cannot say which is the molecule, for there are six ways a given atom, say black, can be paired off with a white.

Third substage—methane series

Identity of the molecule is the next development. This innovation is due to a different kind of bond, the *covalent* bond, which does not depend on attraction. Instead, combining atoms share two electrons, creating a relatively permanent unit. These molecules constitute a sort of alphabet (like the atoms which constitute stage three). Crude oil is like an alphabet soup of these elemental substances which are sorted into what is known as the methane series by the process of refining. Typical molecules of this series consist of one or more carbon atoms combined with hydrogen.

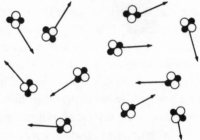

Hydrocarbons: covalent bond
Electrons locked in their molecules

What is important to our philosophic overview is not their contribution to our economy (gasoline, etc.), but that they comprise an ordered set of molecular units which combine with other molecular units to form the great variety of organic compounds of substage four, much as the twenty-six letters combine to form the enormous vocabulary of words.

Additional information

The simplest example of the covalent bond is hydrogen itself, which occurs naturally as molecular hydrogen, H_2, with two hydrogen atoms sharing two orbital electrons.

Hydrogen

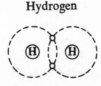

While not a hydrocarbon, because there is no carbon, this molecule could be considered the zeroth member of the hydrocarbon series (see below). It is the simplest covalent bond.

This bond differs from the ionic in giving the molecule an identity which the ionic bond does not have. The ionic bond depends on attraction, and this attraction is felt by the neighboring atoms, resulting in the bonding of many atoms. The covalent bond, like a marriage, binds the partners to one another, making the partnership a unit with relative permanence.

It is this characteristic permanence, or resistance to disassociation, that leads to the term "nonfunctional compounds."

Oils are an example; they do not dissolve in water, and they do not readily combine with other molecules.

The methane series, an important third-substage group, consists of combinations of carbon and hydrogen according to the general formula C_nH_{2n+2}.

Methane	CH_4	H H C H H	A light gas
Ethane	C_2H_6	H H H C C H H H	Gas
Propane	C_3H_8	H H H H C C C H H H H	Cooking gas
Butane	C_4H_{10}	etc.	
Pentane	C_5H_{12}	"	
Hexane	C_6H_{14}	"	Begin gasoline
Nonane	C_9H_{20}	"	Begin kerosene

The series adds more and more units of two hydrogen and one carbon until we get fuel oil, lubricating oil, and finally asphalt. This series is also called the paraffin series. The word "paraffin" is derived from "parum" (too little) plus "affinis" (akin), in allusion to its chemical inactivity.

Like atoms, the members of the methane series derive their properties from their position in the sequence; thus the volatility which is desirable in gasoline, and the inertness of asphalt are properties which depend on position in the sequence.

The series based on the benzene ring is also third-substage. It has the general formula C_nH_{2n-6} and consists of carbons in a hexagonal ring at each vertex of which are hydrogens or hydrocarbons.

Benzene series

Benzene (C₆H₆) Xylene (C₈H₁₀)

Xylene with two methyl groups has three forms depending on whether the methyls are adjacent or separated by one or by two hydrogens.

Other hydrocarbons combine two or more benzene rings, and it is worth noting that the number of positions in a ring follows the same sequence that electrons do in atoms: 2, 6, 10, and 14.

Fourth substage—functional compounds

Combining the members of the methane series (the alphabet) with radicals (other kinds of letters) creates the *functional* compounds. Hundreds of thousands of kinds of organic molecule become possible through this combination of combinations. For example, wood alcohol is made from methane plus the OH group; grain alcohol from ethane plus the OH group; and ketones, esters, and amines are derived from other groups. Other alphabets, such as the benzene series, make many more complex molecules.

Additional information

The fourth substage introduces the complexities of organic chemistry and the endless varieties of compounds. So let us not get in too deep. The compounds of the paraffin series we have just considered are nonfunctional; they do not readily associate with other molecules. But if we replace a hydrogen in one of that series, say ethane, by an OH radical (or group), we obtain an alcohol, ethyl alcohol, which is notorious as a good mixer.

What has happened? Alcohol is soluble in water; its contribution is to reintroduce the ionic bond* and communal attraction. The married couples

---

* In the form of the hydrogen bond, a weak ionic bond. The weakness of the hydrogen bond is one of the factors that makes life possible, because hydrogen combines at room temperature.

meet at a cocktail party and are attracted to one another; all sorts of combinations result.

There are other alcohols. Not only does each of the methane series produce an alcohol—methane produces methyl or wood alcohol; ethane, grain alcohol; propane, propyl alcohol; and so on—but we can have alcohols in which two or more hydrogens are replaced by OH radicals. Such is diethylene glycol, used as antifreeze under the trade name of Prestone.

And there are other radicals. Thus the aldehyde radical, CHO, leads to a whole series of aldehydes, beginning with formaldehyde; the ketone radical, CO, creates the solvents used in lacquers and nail polish. The amines, derived from ammonia (NH_3), are obtained similarly. The whole series of carbolic acids found in soap descends from the substitution of COOH for hydrogen in various members of the methane series.

Even more interesting are the esters which are responsible for the flavors and odors of fruits and flowers. An example is isoamyl acetate, or banana oil, formed from amyl alcohol and acetic acid.

So far we have listed only the compounds which derive from the methane series. Many more, including the sex hormones, derive from the benzene series. This should suffice to describe the character of the fourth substage, all of which are *compounds of compounds*.

We have emphasized especially organic compounds because of their importance to life, but minerals that cannot be classified otherwise, as salts, etc., may also qualify as fourth-substage.

Fifth substage—polymers

What advance could be made beyond this combination of combinations? We find it exemplified in the polymers, which consist of chains of hundreds of thousands of units, each unit being made up of some dozen or more atoms. Cellulose, rubber, rayon, and nylon are all examples of polymers. While the structure of the polymer may seem monotonous, the repeated links in the chain are the evolutionary jump which make possible greater wonders to come. Polymers resemble plants not only in producing a chain of cells, but in their capacity to grow. All chemicals from polymers on share this power.

Additional information

Polymers are *chain* molecules, with hundreds of thousands of identical links. The designation "nonfunctional" refers to their chemical inertness, a consequence of the covalent bond which comes in again here in confirmation of their placement at level III. Another property that marks them as fifth-substage is that they grow, as do cells, in a chain or *series* of links.

The analogy which likens atoms to letters, and molecules to words, could be extended to make plants analogous to sentences. The *life* of plants, like the *meaning* of sentences, is this emergent factor. Applied to polymers, the life-like factor is their ability to store order by drawing energy from the environment. This is an "uphill" manifestation; it goes against the second law of thermodynamics, which decrees that entropy (see beginning of Chapter VIII) is positive. Thus polymers have negative entropy, which is a reminder that the turn has occurred. It is worth noting that the wood of trees consists of polymers, cellulose and lignin, i.e., the fifth stage depends on fifth-substage chemicals.

A conjecture at this writing, still not proved, is that the links of the nonfunctional polymers are fivefold, that is, the number of atoms in a link is a multiple of five.

Sixth substage—proteins

Beyond the polymers in order of complexity are the proteins, which are chains with side chains. The side chains provide means to twist the chain into one of several kinds of helix, and may also hook up to other points on the chain, thus making possible molecules which have an elaborate spatial structure, such as the hemoglobin in blood, which makes a container for picking up oxygen at the lungs and transporting it to the muscles. Blood, hair, nails, skin, horns are all proteins. Thousands of kinds are needed for the animal organism. This multiplicity of shapes and structures of proteins, their mobility and their role in digestion, suggest a resemblance to the sixth or animal kingdom.

We can even see in this protein chain with side chains a foreshadowing of the segmented animal body with its articulated feet —the arthropod.

Additional information

The proteins are the functional polymers in Dr. Price's list at the beginning of this chapter.

With proteins, the world of molecules acquires magical powers. Impressive as is the contribution of polymers to the structure of trees, we have here an even greater marvel, the possibility of animated substance. For example, actin and myosin. These proteins, which constitute muscle, construct organized battalions of molecules interlaced with veins and arteries (to supply fuel and remove waste) and with nerves (to control). Actin and myosin constitute a chemical that moves.

I will never forget the surprise I received when, having anticipated the existence of a mobile chemical, but not really believing that it existed, I was told that there are indeed mobile molecules, and that muscle, constructed molecule by molecule, is a living structure, one that can move by shortening and lengthening itself.

It is also impressive that a molecule can anticipate the mobility we would expect only of animals. And the movement is not the random motion of particles, such as electrons; this is a coordinated movement of billions of molecules. In fact, recognizing that the bulk of the animal is muscle, we may say that the muscle is "the word made flesh," the vital principle serving the will.

Albert Szent-Györgyi discovered that the long actin and myosin molecules overlap their side chains, interlocking like oars of racing shells placed side by side. Activation of the muscle causes the side chains to swing, or *shrink*, in such a way as to draw the actin and myosin together, thus shortening the muscle.

The more inert or *structural* proteins include the keratins: skin, hair, fur, wool, horns, nails, claws, hoofs, scales, beaks, and feathers.

Other proteins, including enzymes, are *functional*. Lysozyme, a chain of 129 amino acids, destroys bacteria by wrapping itself around the molecule in the wall of the bacterium and pulling it apart, a sort of animated monkey wrench like the bewitched broom in "The Sorcerer's Apprentice."

A few proteins have been analyzed. Many years of research were required by Sanger and his collaborators to establish the structure of insulin, which was found to have a molecular weight of 12,000 and to consist of four polypeptide chains, two with twenty-one amino acids and two with thirty. (The amino acids are the side chains referred to above.)

Hemoglobin, which transports oxygen in the blood, is a protein built on a core of four iron atoms. It is estimated that man requires some 50,000 different proteins.

The structure of protein was discovered by Linus Pauling and is called a polypeptide chain; it consists of a chain of links.

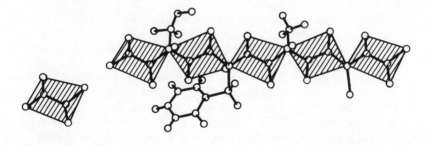

The units of the chain, known as polypeptide links, each consist of six atoms held rigidly in a plane, with carbon atoms at each end joining the links. To each of these joining carbons is attached a side chain, one of twenty amino acids having different structures and different chemical properties. One function of these side chains is to twist the links into a prescribed angle. This causes the chain to form a helix, or, when alternate links are rotated 180°, a ribbon. In both cases, the open corners are joined by hydrogen bonds (a form of ionic bond). There are three possible helices of different pitch, which may be left or right.

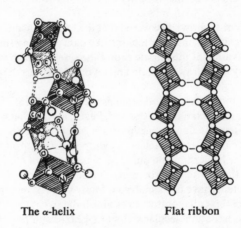

The α-helix   Flat ribbon

The *structural* proteins mentioned, hair, skin, etc., consist of helical chains. Silk consists of the flat ribbon.

The *functional* proteins are more complicated. They include side chains which cause the helix to bend to form a corner at predetermined points, thus making possible molecules such as the hemoglobin of blood, which is a container for the oxygen-carrying iron atom. Other functional proteins such as enzymes make miniature laboratories in which particular molecules can be synthesized or broken down.

Thus the linked peptide units, together with the side chains that determine their angle, comprise a sort of construction set like the mecanno toy which makes possible a huge variety of shapes. Still I find it incredible how even this ingenuity can construct a feather.

To sum up, the functional polymers or proteins at substage six add a much greater versatility to the growth power of the previous substage. This versatility, in the functional polymers especially, becomes a kind of animation, a mobility. Further study is needed, but it appears that the extra range of behavior, or freedom, of the proteins is attributable to the ionic bond which reenters the picture at the sixth substage. Recall that the fifth

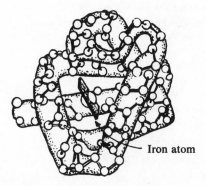

Iron atom

Myoglobin: one-fourth of the hemoglobin molecule which carries oxygen in the bloodstream

substage restores the covalent bond, which holds the links of the chain together. In proteins, it is the ionic bonds of the side chains that provide the versatility of helices and different bonds, not to mention the contraction of the side chains which join actin and myosin. The ionic bond is the sexy one, even as it was at substage two where the sodium atom gives its electron to fill the hole in the chlorine shell.

A final note: according to our arc, the sixth-substage bond should have two degrees of freedom. It is clear from the variety here possible that the bond is much more versatile than in the fifth substage, but I have only recently noted the feature of the peptide links that makes the two degrees of freedom precise. This is that the peptide links can be twisted with respect to one another through *two* angles ϕ and ψ.

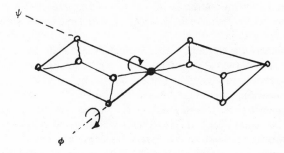

It is these two angles, each a degree of freedom, that make possible the variety of structures and confirm the prediction of two degrees of freedom.

Note too the increasing homogeneity at this substage. All proteins are polypeptide chains, whereas the fifth-substage polymers came in great variety.

Seventh substage—DNA

Recently discovered by Watson and Crick, DNA consists of a double helix whose ladder-like rungs code the information necessary for the cell to duplicate itself or for it to build a multicellular organism. All DNA is of the same external shape in contrast to the different shapes of the proteins it controls. This fact would seem to demonstrate the same principle which may underlie the fact that man has a singular form while animals take so many shapes. As the DNA (in the seventh substage) carries information and manufactures the proteins that do the work, so man (in the seventh stage) makes the tools he needs and uses the animals he domesticates to do his work.

Additional information

Prior to the discovery of DNA in 1953, biologists for some time had recognized that cellular processes, including cell division and inheritance, are governed by particles, visible under the microscope, which they called chromosomes. Watson and Crick made the breakthrough that established that chromosomes ultimately consist of DNA molecules. They found DNA to be made up of a double helix, two chains of phosphates linked like a ladder by rungs joining the two helices together. The rungs are of four different kinds which, taken in groups of three, provide a twenty-letter alphabet, much as dots and dashes make up the Morse code.

Each of the twenty "letters" of this alphabet designates one of the twenty amino acids which make up the proteins discussed in the preceding section on the sixth substage.

Thus the DNA spells out a message which describes the exact sequence of amino acids required for manufacture of the particular proteins which in turn build the organism. While it is not known how the timing of the manufacture of the thousands of necessary proteins is achieved, it is known that the molecular weight of DNA in one of the simplest organisms, *Escherichia coli*, is over 3 billion (3 billion protein–electron pairs). This would allow for at least 6 million letters in the message (allowing for fifty atoms of an average molecular weight of 10 for each rung of the ladder). This amount of information would require a two-thousand-page volume. DNA therefore can be regarded as a kind of blueprint, coding instructions for manufacture.

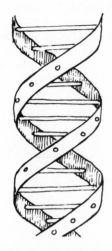

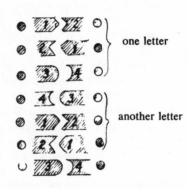

one letter

another letter

It is significant for our theory that the culmination of molecular evolution is a single species of molecule—DNA. We may recall that the protein molecules (substage six), while all built on the polypeptide chain, have many shapes: the strands of fibrous protein for hair, the complex cage of hemoglobin, the interlocking batteries of actin and myosin for muscle. With DNA we have but one shape, the double helix, whose variety is not in its outward form, but in the information it encodes. If we liken the proteins to different kinds of tools, the DNA could be likened to a book—its external shape gives no clue to its content. This uniformity of shape of all DNA molecules is an expression of the homogeneity we referred to in Chapter IV as characterizing photons. This homogeneity is an expression of the top level, the "intelligible" world of light, in this case because of the encoding of pure information. Molecular evolution ultimately reenters the "intelligible" world of light, in this case by encoding pure information.

This brings into focus another fact about DNA, which is that while it encodes the instructions and dictates manufacture, it does not itself do the work; it delegates this function to the so-called messenger RNA which, put together by the DNA, goes out into the cytoplasm (the body of the cell) and manufactures proteins. This aloofness from "manual" exercise helps us to recognize that even here, at the molecular level, there is a distinction between the governing or control principle and the manufacture itself.

We are thus led to recognize the key word "dominion" as suitable for the seventh substage (see p. xxiii).

The symmetry, or lack of it, that we expect at the seventh substage is also borne out, for the double helix, being a spiral, lacks radial symmetry, and, being different along its length, lacks axial symmetry. This lack of symmetry correlates with the complete dedication of DNA to meaning; it is a message with no flourishes (see p. 40).

Because it is made up of DNA, we should include the virus as seventh-substage. Virus, in fact, is almost pure DNA, for it lacks the other substances contained in living cells that make possible reproduction. To obtain these substances, the virus invades the cells of plants or animals and "changes the blueprints" of the host DNA, to force it to make more virus. So the virus is a sort of desperado or highjacker, whose takeover of the cell is a clear-cut example of dominion. In fact, it was the virus that suggested to me the power of the seventh stage in general. While we can condemn its behavior, we owe our existence to the similar capability of our own DNA, for without it, when we eat chicken, we would turn into a chicken.

Let us now take a closer look at the structure of DNA. The cross links or rungs of the ladder consist of four nucleotides which join as pairs; cytosine to guanine and adenine to thymine.

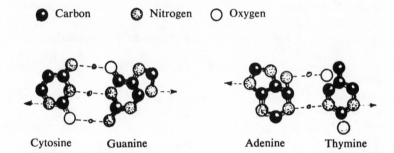

● Carbon   ◉ Nitrogen   ○ Oxygen

Cytosine   Guanine   Adenine   Thymine

Each pair can be reversed end to end, making four possibilities: CG, AT, GC, and TA. The end of each pair joins to a pentose sugar, and this in turn to a phosphorus atom (together with its four oxygens).

When the whole molecule splits, each half retains a complete complement of half units which prescribe their mates on the new chain-to-be, making it possible for each half to duplicate its matching partner and become a new double helix. If you think this is complicated, read Watson's *The Double Helix.** It was his recognition that complementary, not similar, pairs went together that solved the mystery of DNA and led to Watson's receiving the Nobel Prize.

One further point is of speculative interest: counting guanine and adenine, each of which contains a five- and six-ring as two units, and counting along a rung from one end to the other, we find there are seven elements on each rung of the DNA ladder.

* Watson, James. *The Double Helix: Being a Personal Account of the Structure of DNA.* New York: Atheneum, 1968.

While the significance of this is unknown, it fills in the picture of an ascending complexity in the molecules of each substage. (The peptide units were sixfold.)

This kind of conjecture, except for the periodic table itself, is not a part of current chemistry, but is promising as a speculative device. There is certainly an increasing complexity as we go through the levels of organization of molecules; and aside from the mere count of the number of atoms in a molecule, or even of the kinds of atom required, there is the possibility that the numbers from one to seven can distinguish kinds of molecular topology, and therefore indicate a more profound relationship than is recognized in current science. (See Appendix II.)

Correspondences on the arc

To return to the molecular kingdom as a whole, we can now appreciate that the development that has taken place from the metals to DNA is itself a process.

If we arrange these substages in an arc, we discover something of great interest.

Bonds

Level I	Monatomic (Metals)	DNA + Virus	Metallic
Level II	Ionic (Salts)	Functional Polymers	Ionic
Level III	Nonfunctional Compounds	Nonfunctional Polymers	Covalent
Level IV	Functional Compounds		Ionic and Covalent

When we examine the type of bond which prevails in each substage, we find substage four combines ionic and covalent bonds, substages three and five are covalent, substages two and six are ionic. Since the monatomic molecules are metals, substage one is, of course, metallic bond. We *predict* the same for DNA, which seems likely in view of its probably superconductive core. (This suggests the possibility that the DNA radiates a short-wave radio signal which coordinates cell growth.) Less speculative is the *difference in freedom* of the bonds on the successive levels. At level I the bonding electrons of metals move freely through the body of the metal, that is, the bond is unlocalized. At level

II the electrons of ionic compounds create bonds which connect each atom with its neighbors. And at level III the electrons of covalent compounds are confined to the molecule itself.* (See illustration, page 69.)

Correspondences of substages to kingdoms

Not only are there seven substages, but each substage has a similarity to the corresponding kingdom:

1. The *conductivity* of metals, with their free-ranging electrons, corresponds to electromagnetic radiation in the kingdom of light.
2. The *binding* of ionic compounds, with their positively and negatively charged atoms, corresponds to the proton and electron, the positive and negative particles.
3. The *identity* of the hydrocarbons in the paraffin series corresponds to atoms, which also form an "alphabetical" sequence with properties determined by their position therein.
4. The *combinations* which we call functional compounds are fourth-substage because they are combinations of combinations.
5. The polymers in their self-replication and chains of units correspond to plant *growth* with its cell division.
6. The proteins, as the chemicals that *move*, correspond to the animal kingdom.
7. DNA, which *directs and governs* the growth of plants and animals, gives us the clue to the distinctive character of the seventh kingdom; we can now call it the *dominion* kingdom.

Dependence of seventh stages and substages on what is beyond themselves

Note that we have defined the dominion kingdom (seventh stage) from evidence drawn from the seventh substage, DNA and virus.

---

* The free movement of electrons in metals makes possible electrical conductivity. Their confinement (at level III) makes insulation.

Both DNA and virus are boss molecules which direct the activity of the cell "factory" and hence warrant the dominion designation—but there is an interesting additional point—that is: that the seventh substage, DNA and virus, *requires cells* for the completion of its function; and *cells belong in the next higher kingdom.* It will be found that all seventh substages require the next higher kingdom to function. For example, the flowering plants depend on insects for pollination. Equally remarkable is the dependence of controlled radioactivity of atoms (third stage) on molar concentration (fourth stage). It was once thought that radioactive disintegration could not be influenced by any external factor, but the atom bomb is evidence to the contrary.

Both the atom bomb and the atomic pile depend on bringing together enough uranium atoms so that the disintegration products of one atom (neutrons) will cause the disintegration of other atoms. In the bomb this process proceeds instantaneously, causing an explosion. In the pile the process is moderated or slowed down by carbon rods pushed between the atoms to absorb the radiated neutrons and control the rate of fission.

What would this mean to the dominion kingdom? Since it is already the highest kingdom for the solar universe, we are led by this and other evidence to expect yet higher stages—a super arc which deals in galactic evolution. In other words, the dominion kingdom requires something beyond itself, which may help to explain why all human cultures, with the possible exception of modern man, depend on a belief in higher orders of beings, gods.

This dependence of seventh substages on the next higher stage is one of the most difficult concepts to accept because it suggests that process, at the seventh substage at least, anticipates its own future. Despite this seeming incredibility, I think we should try to heed this point, for the evidence presented in the substages is confirmed by the theory. That is, the phenomena suggest a recurrence of the teleology implied by the behavior of light, whose capacity to take the shortest path to its goal so impressed Leibniz and Planck.

Powers of the kingdoms

We designate below the key word or *power* for each kingdom (stage):

Kingdom	Power		Image
1 Light	Potential		Point
2 Nuclear	Binding		Line
3 Atomic	Identity or form		Circle
4 Molecular	Combination (or separation)		Two circles
5 Plant	Growth or organization, reproduction		Chain
6 Animal	Mobility		Chain with side chains
7 (Man)	Dominion		?

The power of the first kingdom is listed above as potential. This designation is not entirely justified by the corresponding molecular substage, but we will show the appropriateness of this choice in later development.

A final point about the molecular kingdom, sufficiently arresting to deserve comment: the kingdom starts by creating the properties we associate with matter—density, hardness, strength, conductivity, boiling point, etc. These are the properties which emerge when millions of atoms are brought together. *As the power of combination develops,* these properties undergo a subtle change. What seems to occur is that the individual molecule acquires a kind of responsibility: it extends its domain more and more, until it is not only itself organized from millions of atoms, but (as DNA) it builds an organism billions, trillions of times its own size.

The grid

On the basis of our survey of the atomic and molecular kingdoms, we can now put forward a scheme which divides each of the seven kingdoms into substages. Each substage contributes, to the kingdom of which it is a part, a thrust or power similar to the thrust or power which the numerically corresponding kingdom makes to the whole of process.

Referring to the grid: the left-hand column lists the kingdoms by number and name. Below each name is the word for the corresponding power, together with other words that characterize the kingdom. In the upper right-hand corner of the rectangle of each kingdom is its degree of freedom and symmetry. The next seven narrow columns describe the substages. Note that the key word for each substage is the same as that for the power of the kingdom.

The top row of the grid divides the light kingdom into substages by frequency (CPS) beginning with the highest frequency (or greatest energy) at substage one. Thereafter each substage begins with a frequency approximately 1/2000 as great. Wave length is shown in cm., and energy in electron volts. So divided, each substage includes the frequencies that activate the corresponding kingdom.

Substage one: Cosmic rays—high energy particles of more than 1 G.E.V.

Substage two: Photons that create protons to electrons, gamma rays

Substage three: X rays, atomic spectra

Substage four: Visible radiation and molecula. spectra

Substage five: Microwaves about 10^{-2} cm. to 10 cm. (predicted for cells)

Substage six: Shortwave radio 100 cm. to 10 meter (predicted for animals)

Substage seven: Long-wave radio 10 meter to 20000 meter

Note that visible light is in the exact center and that room temperature (at which life begins) comes at the beginning of the fifth substage.

The substages of the nuclear kingdom have not yet been assigned in satisfactory fashion and are left blank. The substages of the atomic and molecular kingdoms are set forth as we have discussed them in both Chapter VI and this chapter. The substages of the vegetable and animal kingdoms will be covered in the next two chapters. The last, or dominion, kingdom will be discussed in the final chapters.

We have discussed the atomic and molecular kingdoms. The

KINGDOMS ↓	STAGES →	POTENTIAL	BINDING
1. LIGHT POTENTIAL: No mass, outside of space and time. Quanta of action. Hierarchy.	3 degrees of freedom; no symmetry	10^{25} 10^{-15} 10^{11} **Cosmic rays** **Proton rest energy** ⟶	10^{22} $10^{-1?}$ 10^{7} **Gamma Ray** **Nuclear binding energ**
2. NUCLEAR BINDING: Substance, force. The spell aspect of image, hence illusion. "Probability fog."	2 degrees of freedom; bilateral symmetry		
3. ATOMIC IDENTITY: Acquires its own center. Elements, order creates properties. Exclusion Principle. Rows of Mendeleef Table →	1 degree of freedom; radial symmetry	$\boxed{2}$ **HYDROGEN** **One 2 Ring**	$\boxed{2}\,\boxed{2}$ **LITHIUM to FLUORINE** **Two 2 Rings**
4. MOLECULAR COMBINATION: Molar properties. Classical physics, determinism. The only kingdom we see.	0 degree of freedom; complete symmetry	**METALS** **Single Atoms**	**SALTS** **Double Atom**
5. VEGETABLE GROWTH: Self Multiplication. The cell or organizing unit. Order building by negative entropy.	1 degree of freedom; radial symmetry	**BACTERIA** **One Cell**	**ALGAE** **Many Cells**
6. ANIMAL MOBILITY: Action and satisfaction. Digestion, mobility. Choice becomes possible.	2 degrees of freedom; bilateral symmetry	**PROTOZOA** **One Cell**	**SPONGES** **Many Cells**
7. DOMINION CONSCIOUSNESS: Memory of one's own acts leads to knowledge and control.	3 degrees of freedom; no symmetry	**?**	**TRIBAL SOCIET** **(No Bodies?)** **Collective Unconscious**

IDENTITY	COMBINATION	GROWTH	MOBILITY	DOMINION
10^{18}	10^{15}	10^{11}	10^{8}	10^{4} Hz
10^{-8}	10^{-4}	10^{-1}	10^{3}	10^{6} cm
10^{4}	10^{0}	10^{-3}	10^{-7}	10^{-10} eV
X Rays Atomic Spectra	UV IR Molecular spectra	Microwaves Cellular radiation? ← $h\nu = kT$	TV and radio waves Animal radiations?	Low frequency waves
SODIUM to CHLORINE One 6 Ring	POTASSIUM to BROMINE Two 6 Rings	RUBIDIUM to IODINE One 10 Ring	CESIUM to ASTATINE Two 10 Rings	FRANCIUM One 14 Ring
METHANE SERIES Non-functional Compounds	Functional Compounds	POLYMERS Chains	PROTEINS Chain with Side Chains	DNA AND VIRUSES
BRYOPHYTES Tissue	PSILOPHYTALES Many Tissues	CALAMITES Segmented Larger Size	GYMNOSPERMS Mobility of Seed	ANGIOSPERMS Flowers
COELENTERATES One Organ	MOLLUSKS, etc. Many Organs	ANNELIDS One Chain	ARTHROPODS Chain with Side Chains	CHORDATA
	———— MODERN MAN ————→		CHRIST BUDDHA	
Consciousness	Objective Thought	Creative Genius	Mythical Kings Mazda?	?

vegetable and animal kingdoms will be covered in the next two chapters. The first two kingdoms, light and nuclear particles, are not sufficiently understood to analyze at this time. The last, or dominion, kingdom will be discussed at the end of the book.

Recapitulation and preview of later chapters

The importance of this chapter is that it confirms the thesis of an arc. The development from molecules with but one atom to DNA with billions of atoms is as much an evolution as is the evolution of cellular life from one-celled bacteria to trees and elephants. But because chemistry is a more exact science than is biology, it can provide a basis on which to build; it can give sanction to the conjectures on which we based the arc and elevate them to the principles we need to discover the nature of human evolution.

For example, we noted that the evolution from photons to molecules entailed a loss of freedom followed by its recovery under control. This is true within the molecular kingdom, where the bonding electrons in metals are free to move at the speed of light but become increasingly constrained in salts and completely so in oils, ultimately to regain their freedom in proteins and DNA, where controlled electrons make possible the functions of these highly organized molecules.

Again, we saw how the symmetry of the arc of kingdoms (Chapter IV) manifested for molecules in the similarity of bonds, i.e., ionic bond at substages two and six and covalent at substages three and five.

Most interesting is the loss and subsequent regaining of homogeneity —which is dramatically demonstrated in the progression from polymers with many kinds of chains, to proteins using only the polypeptide chain, and again from the variety of side chains in proteins to four of DNA.

This will be important when we come to man's evolution, for the fact that molecules develop from a sixth substage having a variety of shapes to a seventh having one shape (the double helix) is evidence that the contrast between the variety of shapes in animals and the unity of man's shape (all men are of one species) can be seen as a principle.

This, of course, is only part of the story. Despite anatomical similarity, humans are not alike any more than are DNA molecules, or books. Humans are different in their purposes and character rather than their physical attributes. This has importance for human evolution, which we will take up at the end of the book.

But before treating this important theme, human evolution, we need to explore and describe the two types of evolution which precede the human, the evolution of the cellular organism in the plant kingdom and the evolution of the animals; these are distinct from human evolution. The next chapters will be devoted to this subject.

VIII | The plant kingdom

Evolution versus involution

Evolution begins with the plant kingdom. Or perhaps we should say visible evolution, because the very first evolutionary activity comes earlier, with the polymers, which constitute the fifth substage of the molecular kingdom (significantly, the substage just to the right of the low point or "turn" of the arc). The polymers move against the law of entropy. And what is that?

Entropy is the tendency of the energies associated with inorganic substances to become more uniformly distributed—of stones to roll downhill, and hot objects, emitting heat, to grow cooler—so that the total energy in a given area or system gradually becomes unavailable by averaging out. This tendency, described in physics as the second law of thermodynamics, is implied in the so-called billiard ball hypothesis, which conceives of the universe as a gradually subsiding agitation of lifeless objects.

To get back to the polymers: in their self-replication they go counter to entropy; they *store up* energy in the substances they build. That storing up, known as *negative* entropy, anticipates the upward thrust that is evident everywhere in the plant kingdom.

This general thrust in plant life manifests in several ways which we should think of as interrelated:

Negative entropy, or storage of energy
Hierarchy
Increase of order or organization
One degree of freedom: radial symmetry

Size and growth
Self-reproduction
Evolution

But if this is the beginning of evolution, how are we to characterize the development of atoms and molecules? These too are stages of organization and as important to process as are its later stages, and they could also be called evolutionary. However, it is traditional to refer to these early stages as *involution*, and this word conveys the distinctive character of the left-hand side of the arc, in which process is descending, i.e., becoming more *involved* in matter and hence more constrained, whereas on the right-hand side it is moving *up* into higher forms which are more *evolved* and free.

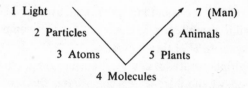

1 Light   7 (Man)
    2 Particles   6 Animals
        3 Atoms   5 Plants
            4 Molecules

In Chapter IV we described this arc with emphasis on the increasing freedom of the stages beyond the molecular. Plants at the fifth stage gain a degree of freedom which manifests not only in their growth but in their ability, through their progeny, to conquer time. Both the growth of the plant (its storage of energy, its organization) and its self-propagation through progeny result from cell division; if the cells remain together we call it growth, if they split off into a separate entity we call it progeny. But cell division is an enormously complicated process.

Complexities of the cell

It might be thought that cell division is merely an increase in size with an imposed limit which enforces division, much as water leaking from a spigot forms into drops of a certain size. But it is becoming apparent that cell division is a very complicated process. It used to be said that the cell is filled with protoplasm, but this description is as inappropriate as to say that a radio is filled with radioplasm.

The cell is the site of a vast number of complicated chemical reactions. These reactions are conducted by enzymes, specialized protein molecules that cause certain chemical reactions to occur about a million times faster than they would occur without the enzymes. The cell also includes DNA, which carries the information, and RNA, which conducts the manufacture of material for cell growth.

Escherichia coli, the bacillus which inhabits the intestine and which divides every twenty minutes, is one of the smallest life forms and the one biologists have studied most thoroughly. It is estimated to contain —in addition to 4 DNA boss molecules of a MW (molecular weight) of 2.5 billion each—about 400,000 RNA molecules (of one thousand kinds) with an average MW of 2 million, about 1 million protein molecules (of two thousand kinds) with an average MW of 40,000, plus 500 million smaller organic molecules (of seven hundred kinds) with an average MW of three hundred.

Such descriptions are, of course, mind-boggling; but to keep a sense of proportion, we should at least recognize that this is of *a different order of organization* from that of even the most complex molecule *per se*. In fact, if we were to liken a molecule to an automobile (both are combinations of different parts), the cell would be comparable to a manufacturing plant, a vast organization of men, machines and computers.

Multicellular organization

Staggering as is this achievement, it is only the beginning of the organization stage of process, i.e., the plant kingdom. The plant is able to organize a multicellular entity—say a tree—which contains trillions of cells whose cooperative interaction is essential to the organism as a whole.

How are we to deal with this kingdom? Should we turn to the biochemist, who has discovered the structure of DNA and the steps by which it conducts the internal chemistry of the cell? If we do no more than that, we will miss something which the lady who says, "My roses are happy today," has not lost touch with—the existence of the plant as a living creature. We must recognize that science can tell us of only a small part of the total picture, for the miracle of plant growth entails the development of the organization principle, and that goes far beyond the chemistry of cell division. This principle is able to guide the growth of the plant, which means that there is something which organizes the plant and is a fellow creature.

I have suggested that DNA with its possibly superconductive core, its coiling, its necessary inductance—together with the fact that all nearby cells have this same DNA—could broadcast at frequencies in the lower part of the infrared, not only to monitor the activity within its own cell, but to coordinate growth steps of neighboring cells. While this is only a supposition, it helps to delineate the problem. Something of the nature of radio broadcast is necessary to account for the coordination of cell growth.

Other principles

Meanwhile, let us proceed and keep our eyes open. Certain self-evident facts about plants, passed over by biologists, are important, such as the fact that the tree is not *all* alive: its wooden trunk is built up year by year by the cambium layer, which is the only *living* part of the trunk. Similarly, while the leaves are alive, the branches that support them are not. Thus the life of the tree is a thin surface film which covers the trunk and branches, in contrast to the animal, whose life penetrates its whole volume. This distinction, like the difference in their degrees of freedom, helps to give us an overall view of the difference between plants and animals.

Another important fact is that it is the growth principle that makes evolution possible. Current science regards evolution as explained by a process of selection, "survival of the fittest." But it does not tell us how something arises to be selected. Selection is a cutting-off process, like getting one's income from cutting coupons, but cutting off does not explain the growth that provides something to cut off. Growth in plants traces back to cell division by which both the mature plant and its reproduction come about. This growth is what is called exponential: it not only increases, but if not checked, it increases at an increasingly rapid rate.

A single bacterium like the *Escherichia coli*, dividing every twenty minutes, would fill a football stadium with its progeny in twenty-four hours if conditions permitted. It is such expansion that makes it possible for selection to work in order to provide a population adapted to survive in a given milieu. This expansive tendency is a major contribution to evolution.

The fork: self-limitation

This leads to another and very subtle aspect of the plant kingdom. The multicellular organism—say, a tree—is not wholly committed to self-expansion, but has the alternative of producing progeny. In this connection, it is interesting that if we wish to make a tree bear fruit, we inhibit its growth, by pruning the roots or branches or even by banding the tree to reduce the nourishment or circulation through the cambium layer. The tree takes this signal as a warning and turns its energies to producing seed. This *choice*, or fork, is intrinsic to the fifth stage and is an integral part of process. We may interpret it as the inverse of that which occurs in the third—self-determination. The inverse is self-surrender. Or if we read the third stage as "taking on a center," the fifth would be giving off centers, by production of seed.

Here we have a most far-reaching application of the deductive tools provided by the arc. When we read "giving up self-expansion" as the inverse of "self-determination," we might seem to be making a mere play on words. But by realizing that the one describes the creation of a nucleus, such as makes the atom possible, and the other describes the creation of seeds, which makes reproduction possible, we can appreciate that our method highlights the most conspicuous contributions of two kingdoms.

The why of the third stage: acquiring identity

But it might be asked: how can we show that self-determination, or "having its own center," and not some other manifestation, occurs at the third stage? How is this principle imbedded in the formalism? This takes us back to the degrees of freedom and constraint as set forth in the arc:

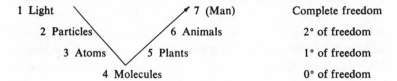

1 Light	7 (Man)	Complete freedom
2 Particles	6 Animals	2° of freedom
3 Atoms	5 Plants	1° of freedom
4 Molecules		0° of freedom

It will be recalled that the second level introduces time, and the creation of nuclear particles having substance but no identity. This lack of identity is due to the fact that there is only one constraint (time). Without another dimension, there is no way to locate or define identity. This definition is the function of the third level, which provides two constraints (dimensions) and thus a basis for definition (limitation) of the endlessness.

The atom, a typical entity, acquires an identity because it possesses a nucleus, an unchanging center, which survives the encounters that affect its periphery (its shell electrons).

The why of the fifth stage: giving up identity

In stage five we find just the reverse: the plant eventually *gives up* the power of unlimited self-increase, its centered negative entropy. (This sacrifice is not an evolutionary setback, for it makes possible the yet greater freedom of mobility or animation that we will take up in the next chapter.) The development of the power which leads to reproduction, and the plant kingdom, reaches its ultimate expression in the flowering plants, or angiosperms, whose perfection of the seed principle has produced the biosphere—the vegetation which covers the earth: grassland, forest, jungle.

Homogeneity of cells

One more important point about plants is that every cell of a plant is a twin brother of every other cell. The tree is a family of cells descended from one. The relationship of the cells is rooted in their DNA, so that they all have the same directive within them. Organizations of cells thus have a more cohesive origin than organizations of people. In the latter, there is always the possibility of conflict. Within a multicellular organism this cannot arise. The function which differentiates a leaf from a root, and which operates through enzymes, has its origin not in the cells, but in the needs of the organism. Apart from such functional differences, all cells of a plant are exact duplicates of one another. Their gene structure is identical. Due to cross fertilization, this identity is not present in the progeny, which have their own identity.

Substages of the plant kingdom

Having divided the atomic and molecular kingdoms into seven substages, we may now do the same for the plant kingdom.

Plant classification

But is there one plant kingdom? Oddly enough, two process philosophers, A. N. Whitehead and Oliver Reiser, who divide the great chain of being into about seven stages, agree in assigning unicellular and multicellular organisms to different *kingdoms*, not just different substages. Their attitude seems to gain support from the kingdoms already covered: protons and electrons when organized make atoms; atoms combine to form molecules; molecules to form cells. In each case, a plurality of units at one level is a unit at the next. So why not one kingdom for cells and another for multicellular organisms?

My answer to these philosophers is that cells introduce a higher type of organization than is involved in atoms and molecules. This new principle, based on cell division or self-multiplication, makes it possible for a single cell to create the multicellular organism *out of itself*. Cells are not like bricks of a house, distinct from the organism they create; they are not *brought together* as are atoms to make a molecule. Thus the plant kingdom, which develops the power of organization and does so through *cell division*, cannot be divided into two kingdoms.

Biologists go to the other extreme and do not even regard the distinction between uni- and multicellular as a basis for classification. They divide algae (seaweed) into seven or eight grand divisions irrespective of the uni/multi distinction. Some algae (euglena and diatoms) are unicellular, and others include both unicellular and multicellular forms.

As may be seen in the accompanying chart, there is considerable difference in systems of classification. The 1942 division shown uses seven phyla (the phylum is a major division) on algae, three on fungi, one on moss, and then puts all higher plants in one phylum! This has always seemed absurd to me, and I am glad to see that Bold, from whose book *The Plant Kingdom* the table is taken, adds nine major divisions to cover higher plants. But while the twenty-four divisions so obtained are an improvement, they are not much help in our problem of levels of organization.

Here it is interesting to note, however, that when Bold discusses the development of the plant kingdom, his chapters follow closely the division we propose concerning that development.

> Our discussion of the diversity of the plant kingdom has included reference to such groups of organisms as algae, fungi, mosses . . . gymnosperms and angiosperms. Could we have started as well at the end of the series or perhaps with any intermediate group and proceeded in a different order? The order of our study is indeed significant for it reflects a series of organisms of increasing complexity.*

Clearly, Bold feels the importance of the levels of increasing complexity which we are stressing, so let us proceed.

* From Bold, H. *The Plant Kingdom*. Englewood Cliffs, N.J.: Prentice-Hall, 1964.

A comparative summary of some classifications of the plant kingdom*

The arrows indicate the fate of taxa in successively more modern systems of classification. When the name of a group is used later at a higher rank, as, for example, in the change from Chlorophyceae to Chlorophyta, the name of the lower group usually is retained as a subsidiary under the higher. The figures in parentheses are estimates of numbers of species.†

Eichler, 1883 (and modifications)	Tippo, 1942	Bold, 1956	Common name	Approx. no. of species
PLANT KINGDOM	PLANT KINGDOM	PLANT KINGDOM		
A. CRYPTOGAMAE	Abandoned	Abandoned		
DIVISION 1. THALLOPHYTA	SUBKINGDOM THALLOPHYTA			
Class 1. Algae	Abandoned			
Cyanophyceae	PHYLUM 1. CYANOPHYTA	DIVISION 1. CYANOPHYTA		
Chlorophyceae	PHYLUM 2. CHLOROPHYTA	DIVISION 2. CHLOROPHYTA		
	PHYLUM 3. EUGLENOPHYTA	DIVISION 3. EUGLENOPHYTA		
		DIVISION 4. CHAROPHYTA	Algae	(19,000)
Phaeophyceae	PHYLUM 4. PHAEOPHYTA	DIVISION 5. PHAEOPHYTA		
Rhodophyceae	PHYLUM 5. RHODOPHYTA	DIVISION 6. RHODOPHYTA		
Diatomeae	PHYLUM 6. CHRYSOPHYTA	DIVISION 7. CHRYSOPHYTA		
	PHYLUM 7. PYRROPHYTA	DIVISION 8. PYRROPHYTA		
Class 2. Fungi	Abandoned	DIVISION 9. SCHIZOMYCOTA		
Schizomycetes	PHYLUM 8. SCHIZOMYCOPHYTA	DIVISION 10. MYXOMYCOTA		
	PHYLUM 9. MYXOMYCOPHYTA	Abandoned		
Eumycetes	PHYLUM 10. EUMYCOPHYTA	DIVISION 11. PHYCOMYCOTA	Fungi (*Sensu lato*)	(42,000)
	Class 1. Phycomycetes	DIVISION 12. ASCOMYCOTA		
Lichens	Class 2. Ascomycetes	DIVISION 13. BASIDIOMYCOTA		
	Class 3. Basidiomycetes			
DIVISION 2. BRYOPHYTA	PHYLUM 11. BRYOPHYTA	DIVISION 14. HEPATOPHYTA	Liverworts	(9,000)
Class 1. Hepaticae	Class 1. Hepaticae	DIVISION 15. BRYOPHYTA	Mosses	(14,000)
Class 2. Musci	Class 2. Musci			
DIVISION 3. PTERIDOPHYTA	Abandoned	Abandoned		
	PHYLUM 12. TRACHEOPHYTA	DIVISION 16. PSILOPHYTA	Psilophytes	(4)
Class 1. Lycopodinae	Subphylum 1. Psilopsida	DIVISION 17. MICHROPHYLLOPHYTA	Club mosses	(1,000)
	Subphylum 2. Lycopsida			
Class 2. Equisetinae	Subphylum 3. Sphenopsida	DIVISION 18. ARTHROPHYTA	Horsetails & sphenopsids	(25)
Class 3. Filicinae	Subphylum 4. Pteropsida	Abandoned	Ferns	(9,500)
	Class 1. Filicinae	DIVISION 19. PTEROPHYTA		
B. PHANEROGAMAE	Abandoned			
DIVISION 4. SPERMATOPHYTA	Abandoned	Abandoned		
Class 1. Gymnospermae	Subclass 2. Gymnospermae	DIVISION 20. CYCADOPHYTA	Cycads	(100)
	Subclass 1. Cycadophytae	DIVISION 21. GINKGOPHYTA	Maidenhair tree (ginkgo)	(1)
	Subclass 2. Coniferophytae	DIVISION 22. CONIFEROPHYTA	Conifers	(550)
		DIVISION 23. GNETOPHYTA	(No common, inclusive name)	
Class 2. Angiospermae	Class 3. Angiospermae	DIVISION 24. ANTHOPHYTA	Flowering plants	(250,000) (71)
		Approximate total		(350,000)

* From Bold, H. *The Plant Kingdom.* Englewood Cliffs, N.J.: Prentice-Hall, 1964.

† Only groups with currently living plants are included.

First substage

Despite the fact that biologists attach so little importance to the uni/multicellular distinction, we must stress that unicellular plants are the first step toward the complexity of which Bold speaks. By definition, they cannot have cell differentiation, and thus they qualify as first in order of complexity. Here we have bacteria and diatoms, all microscopic.

Substage one: bacteria, one-celled

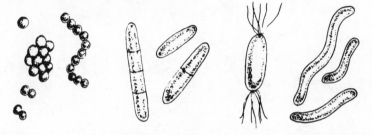

Second substage

The conspicuous development here is a kind of binding that gives rise to great size. Millions of cells adhere together to form a tissue—a plant as we know it, even though a simple one. Here we have seaweed,

Substage two: algae, many-celled (from Villee, C. A., *Biology*, Philadelphia: W. B. Saunders Co., 1972)

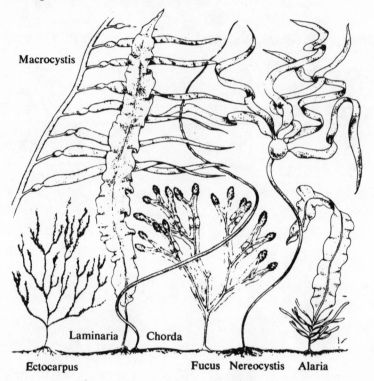

Macrocystis

Laminaria   Chorda

Ectocarpus   Fucus Nereocystis Alaria

including the giant kelp which grows over a hundred feet long. Cell differentiation also has begun; there are stem, leaves, and the holdfast cells by which the plant is attached to the ocean floor. We also find the differentiation of cells required for sex. Instead of two similar cells conjugating, egg and sperm are differentiated, differing in size and number. And the sperm cells have to swim to the egg.

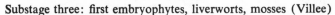

Substage three: first embryophytes, liverworts, mosses (Villee)

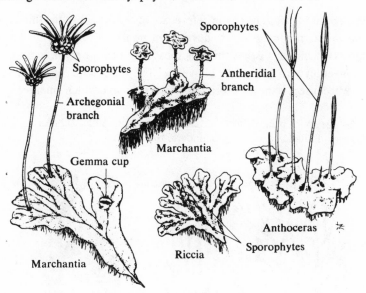

Third substage

All biologists agree that the next development is the mosses and liverworts, the first land plants. These are Embryophyta ("bearing embryos"). They are the first plants to differentiate the embryo from the rest of the plant by forming a *chamber of cells*, with the young plant (itself a multicellular organism) within it.

This is a bull's eye for process theory, because here we have *identity* expressed in terms of reproduction. The embryo is given an identity.

All higher plants are Embryophyta, which is to be expected in view of the cumulative nature of process, each stage or substage

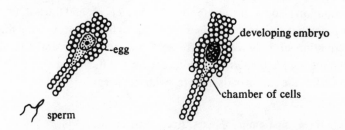

incorporating what has gone before. So, in order to limit the third substage to the first plants to develop an embryo, we should designate it as Bryophyta. "Bryos" means moss and has no etymological connection with "embryo." I have used the word Embryophyta in the grid because it is more descriptive of the principle which emerges here.

Substage four: psilophytes, vascular tissue (from Villee, C. A., *Biology*, Philadelphia: W. B. Saunders, 1972)

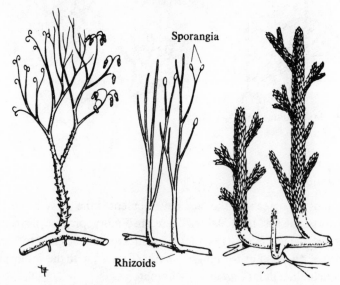

Fourth substage

The next development in plants is vascular tissue, the xylem and phloem which conduct fluid and food through the plant and thus make possible an increase in size. This substage would include psilophytes and club mosses.

Substage five: calamites (horsetails), segmented vascular tissue (Villee)

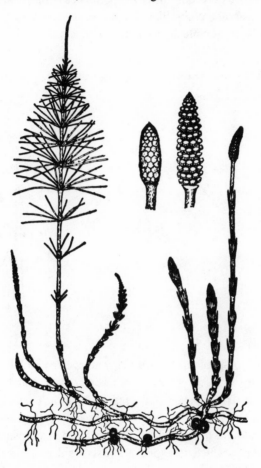

Fifth substage

At this point there is a most important development insufficiently stressed by Bold: segmented vascular tissue, which is the innovation that makes the great difference in size between club mosses and trees. The plants that exhibit this innovation are the Equisetineae or horsetails, called Arthrophyta by Bold and Calamites or Sphenopsida by others. Currently, these are not important plants, but they introduce the production of supportive tissue, and in Carboniferous times reached a height of ninety feet. The strength of modern trees is due to segmented tissue in the form of wood. The correspondence here is that the fifth substage in general is chain-like. Examples: the chains of cells in all plants, the chains of molecular units in polymers.

Substage six: gymnosperms (cycad, ginkgo, red pine)

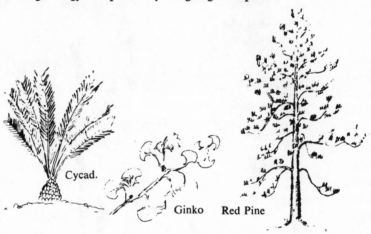

Cycad.

Ginko   Red Pine

Sixth substage

We now come to the plants known as gymnosperms, which include the conifers—pine, spruce, cypress, hemlock, etc.—and which represent, we believe, the sixth substage of the plant kingdom. These are the first of the seed-bearing plants. (The term "gymnosperm" means that the seeds of these plants are naked, i.e., not enclosed in an ovary.) And the factor of *mobility* which we expect at the sixth substage is exhibited in the mobility of the seeds.

Digression on purposiveness

It would appear that, besides growth and reproduction, there are two problems: the distribution of the plant over wide areas and the crossing of genes with other plants of like species. The former is achieved by the transport of spores or seeds, and the latter by having the eggs of a given plant fertilized by sperm from a separate plant, as when pollen is carried from one flower to another.

But in the more primitive plants, the sperm cannot reach a different plant; it has to swim over the surface of the gametophyte (as in mosses and ferns). One might think that these would be placed close together for the more certain achievement of fertilization. But no, in moss the sperm must swim to the top of a long pointed archegonium, and in the fern the sperm is produced at the opposite end of the gametophyte and has to swim its entire length. Why? This question is not answered or even raised in the texts I've consulted, but I think it is important because it illustrates the *purposiveness* which we have referred to earlier as an important "category of explanation" (Whitehead's phrase), and one that has a role throughout evolution.

To repeat the argument of Chapter II, action ML^2/T, is a recognized measure of science, the integral of energy over time (Action $= E \times T$). It was given importance by Planck's discovery that action occurs in *wholes* or quanta that are as fundamental as the proton and electron, if not more so. This wholeness, in conjunction with the fact that action includes mass, length, and time as "parts," enables us to say that M, L, *and T are derived from action, the whole.* Purpose, which of necessity cannot exist in the part, can however be considered to exist for the whole. Indeed, it is synonymous with the function which the whole, and only the whole, can exhibit. (See end of Chapter II.)

In the plant kingdom we encounter a reassertion of this purposiveness that has been dormant in the atom and molecule. The sperm, having the initiatory function in reproduction, can be correlated to the photon, or to first cause, in the life cycle. Like the photon, it is action before its manifestation in matter. Thus the journey of the sperm to the egg is the purposive stage of process; and I believe the separation of sperm and egg is *an evolutionary device to ensure that purposiveness rather than accident start each new generation.* It would be perfectly simple to have the sperm produced very close to the egg, or even in the same compartment, but then accident would provide fertilization with no intention. It is necessary to have a distance between the sperm and the egg in order for the teleological factor to manifest.

How can this be done without reducing the likelihood of fertilization? If it did reduce that likelihood, the progeny would be reduced and the species would suffer and perhaps die out. The answer solves another question that biologists neglect: why are so many sperm produced that one egg may be fertilized? It is by an enormous increase in the number of sperm, and a vast decrease in the likelihood of any given sperm reaching the egg, that the teleological factor is enabled to assert itself.

It could be objected that this is merely a question of the survival of the fittest sperm. But we must recognize that since the sperm is only one cell, its survival cannot be based on the multicellular structure encoded in its DNA, because this structure is as yet unmanifest, and therefore not helpful. The sperm's survival depends, rather, on its liveliness, or on its persistence in moving toward the egg, i.e., on its purposiveness.

Substage seven: angiosperms (daisy, ivy, shagbark hickory)

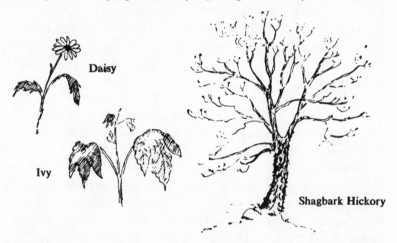

Seventh substage

The seed-bearing plants known as angiosperms constitute the seventh substage of the plant kingdom. The angiosperm ("vessel-seed") is distinguished from the gymnosperm ("naked-seed") in that the seed has a covering. This is the crowning touch that gives the angiosperms their dominance of the plant kingdom. Of the 350,000 species of all plants, 250,000, or 5/7, are angiosperms. Of land plants, they represent an even greater portion. All flowers, grasses, and hardwood trees are angiosperms. The covering for the seed makes possible a variety of devices which aid in seed transport. Such are thistles, burrs, maple samaras which autorotate. So too, nuts, grains, and fruit which animals and birds transport and thereby distribute. Man's main food supply is either directly dependent on such seed covering as in the case of wheat, corn, grains, and vegetables, or indirectly in the case of grazing animals.

Tissue layers

While our division of the plant kingdom is less clear-cut than that of the atomic kingdom, where the relative simplicity of the periodic table gives us a clear structural example of sevenfoldness, or of the molecular kingdom where the powers of binding, identity, combination, growth, mobility, and dominion find expression in the division of molecules supplied by Dr. Price, nevertheless the proposed division at least furnishes an overall survey of the plant kingdom. Even if not ultimately correct, it can serve as a basis for a more penetrating examination of the principles of organization.

What, then, can we learn from this division of the plant kingdom into seven substages? A hint comes from the fact that the angiosperms, or flowering plants, show in cross section *seven* layers of tissue:

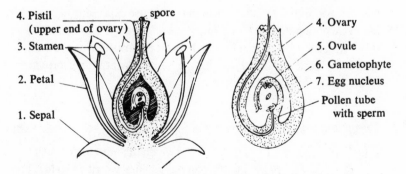

Like the seventh-row atoms, the flowering plant has seven layers or "shells"

The flower is surrounded by an outer layer of sepals (1) which enclose a second layer of more conspicuous petals (2), within which are the stamens (3), source of the pollen, and within that the pistil, which is at the upper end of the ovary (4). Within the ovary are a number of ovules (5), within which is the gametophyte (egg-bearing) (6). Inside of this are eight nuclei, one of which becomes the egg (7), and is fertilized by a nucleus from the pollen microspore. (The process is complemented by a second sperm nucleus in the pollen microspore which fertilizes two other eggs to provide the coating for the seed.)

Without laboring the possibility that this is a valid seven-ness, let us note that there are resemblances to the atom, for the angiosperm in having seven layers is analogous to the complex atoms with seven shells of electrons. There might even be a correspondence in the two nuclei of the microspore and eight of the gametophyte with the 2- and 8-shells of electrons. But if this is true of the angiosperms, how about the rest of the plants? It is a project that would require microscopic examination. However, inasmuch as the sixth-substage gymnosperms ("naked-seeds") have by definition one *less* layer than angiosperms—and since unicellular plants have by definition only one layer—we could say that insofar as evidence is available a correspondence between the number of layers and the substages is borne out.

The number of layers of tissue may, in any case, be a manifestation of a deductive principle and hence a subject for study, and possibly a basis for classification. Recall the Pauli exclusion principle, which so precisely accounts for the seven possible shells of atoms. The layers have an additional significance in that they also represent *stages* in the life cycle of plants. Thus the plant starts as a seed, it grows, it blossoms, it is fertilized, it fruits, and finally the fruit decays, leaving the seed which still contains the germ plus a coating of starch that provides nourishment for the infant plant.

Fungi and retrogressive stages

Fungi, an important division of plants which includes molds, lichens, and mushrooms, are not included in the scheme described so far. This omission would be a serious objection to the sevenfold division of the plant kingdom were it not that the other kingdoms include entities that do not follow the pattern. What is unique about fungi is that they have no chlorophyll and do not engage in photosynthesis, which means that they do not, like all other plants, draw energy from sunlight and store order against the flow of entropy; they are *not moving up* the arc.

Similarly, there are atoms which do not naturally combine with other atoms. These are the noble gases: helium, argon, neon, krypton, xenon, and radon. As for particles, there is the so-called antimatter, which reverts to radiation. The animal kingdom too has its "won't play" entities, the tunicates, sessile animals, which are actually Chordata, the highest subdivision, but resign their opportunity to use mobility and revert to a sponge-like existence.

To complete the scheme, we would expect molecular entities that remain in limbo and do not "progress." This category is actually most obvious of all: the molecules that constitute minerals, which take no active part in evolution.

The addition of these "whiskers" to the arc does not change the sevenfold nature of process. It does indicate the continual presence of option in that, at each stage, there is a choice which, if not made correctly, results in a renunciation of evolution.

IX | The animal kingdom

If the implications of the organization principle in the plant kingdom have stretched our minds, those of the animal kingdom are even more demanding.

Characteristic powers of animals

Voluntary movement

The animal, of course, incorporates the power of organization and growth which it inherits from the plant kingdom. But it develops a new power. The goal of the animal kingdom is *mobility*. This is not mere motion, nor even self-motion like that of an automobile. It is *voluntary* motion. The animal moves in quest of food. For the grazing animals the quest is continuous; for predators, occasional but more strenuous. And almost all animals are under constant threat from natural enemies. The animal therefore requires sense awareness to find its food and to warn it of enemy approach. It must evaluate conflicting stimuli and must choose between alternatives. The deer may be motivated by thirst to go to the water hole, but if it scents a lion it will refrain. The animal is perpetually having to decide on a course of action, when and where to move. This is a different order of mechanism from the tropism of plants. In fact, it is not a mechanism at all; it involves *value judgments*.

This is not true of the "movement" of plants toward light. A potato sending out shoots in a cellar with two windows does not choose one window; it moves toward maximum light and no choice is involved. Even if it be insisted that the animal is moved by the stronger stimulus, this interpretation cannot account for the fact that various kinds of sense data (taste, sight, hearing, smell), as well as internal feelings of hunger, fear, cold, have to be evaluated by a central command, itself not sensory, to lead to action.

We may indicate the difference between the plant tropism and animal motivation by a diagram:

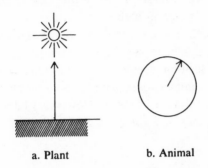

a. Plant   b. Animal

The plant growth is movement toward an *indefinitely* remote goal, which is in a *constant direction*, either against gravity or toward light. The animal movement is toward specific and *attainable* goals—water, food, a mate, shelter—and *in any direction*. This simple distinction between one dimension (the line) and two dimensions (the "radius" of action) was described in Chapter IV, where we showed that animals and plants, like particles and atoms, exchange freedoms and constraints.

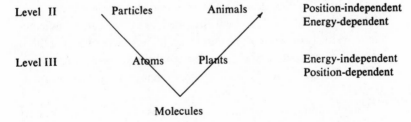

Level II	Particles	Animals	Position-independent Energy-dependent
Level III	Atoms	Plants	Energy-independent Position-dependent
		Molecules	

It will be recalled that we were able to show that the uncertainty of position and of velocity (whose product cannot be less than a certain value) is a property common to both nuclear particles and to animals, on level II, and that the uncertainty of contained energy is a property of both atoms and plants on level III.

Flexible shape

This is the deductive basis which prescribes the nature of the powers, quite apart from their development in plants and animals. Emphasis on this deductive principle guides us in further delineation of their character. For example, we have stressed that identity depends on form and appears at level III. Thus the atom has a characteristic form. For example, it is carbon with four valence electrons and it forms tetrahedral crystals. Similarly, the oak has a fixed manner of growth, a characteristic leaf, and so on. The animal, at level II, has a shape even more precise in some respects than the plant because its anatomy involves a precise number of bones and because it has a characteristic covering of scales, fur, or whatever. However, while the plant shape is fixed, the animal's is flexible. In fact, it is due to this that it is able to move.

Let us now relate this change of shape to the exercise of choice, for while clearly choice and voluntary motion have a practical relationship, it is worthwhile to see the theoretical interconnection. Shape in the sense of a fixed outline is two-dimensional. Even a three-dimensional object can be formulated or described by two or more two-dimensional views. Value, by contrast, cannot be formulated, but value can be indicated comparatively on a one-dimensional scale; it is more or less of this or that. (Weight, cost, temperature, for example, are one-dimensional.) We can therefore say the *shape* of a plant has two constraints and may be *form*ulated.

Such formulation has indeed been proposed. Not only do the curves of plant stems, leaves, and blades of grass follow exponential spirals, but the development of cellular growth in flowers and pine cones proceeds according to an exact formulation, known as the Fibonacci series,* named after the great Italian mathematician, Leonardo of Pisa.

* Examples of the Fibonacci series in nature can be found in Peter S. Stevens, *Patterns in Nature,* Boston: Little, Brown and Co., 1974.

This formulation states that the terms of the series be such that each term (*a*) is to its successor (*b*) as the successor (*b*) is to their sum (*a* + *b*), i.e., $a:b = b:(a + b)$. The ratio *a/b*, known as the golden mean or ϕ, is equal to $(1 + \sqrt{5})/2 = 1.618$. The series of whole numbers $0, 1, 1, 2, 3, 5, 8, 13, 21, 34, 55 \ldots$ obeys the rule that each term is equal to the sum of the two preceding terms, and the ratio of any two terms approaches ϕ as a limit.

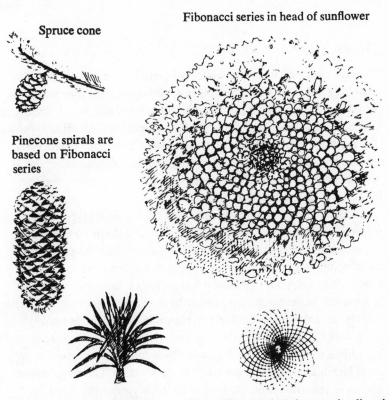

Spruce cone

Fibonacci series in head of sunflower

Pinecone spirals are based on Fibonacci series

Spiral curves in plant leaves

The sunflower spirals in opposite directions: 21 to right and 34 to left

The sunflower shown in the illustration, freely adapted from an informative article by Martin Gardner in *Scientific American* (March, 1969, p. 118), is an example of a flower based on two sets of spirals, 21 to the right and 34 to the left, numbers which are successive terms in the Fibonacci series. According to Gardner, giant sunflowers have been developed that go as high as 89 and 144 spirals.

It will be noted that the formula for the golden mean involves the square root of 5. To imply a connection between the number 5 and the fifth principle might incur skepticism, but in Appendix II we show that the joining of n points by lines of equal length is not possible unless $n < 5$. A corollary is that five points are required to store energy (as in a spring), so that the connection of five-ness with growth is profound.*

Value scale for animals

The *value* scale of an animal has one dimension and can be measured: Fido likes bones best, dog biscuits less, and so on. This may seem a bit trivial, but it is not, for this statement illustrates the reduced constraint and increased freedom of the animal as against the plant. It also shows that there are two mutually exclusive kinds of meaning: the formal meaning of definition and formulation, and the value meaning that is plus or minus on a scale. The former, level III, is objective, and the latter, level II, is subjective.

The shift that occurs between the plant and animal stages of process is just this shift from a formulated blueprint of growth (the cellular organization as prescribed by DNA) to a value scale (as prescribed by inherited instincts) ranging from positive (pleasure) to negative (pain). The behavior of an army undergoes a similar shift when war is declared. In peacetime the army activity is completely taken up with rules of organization as set forth in manuals—drills, code books, procedures. In time of war the emphasis shifts to defending or capturing a position, to winning a battle, goals which cannot be achieved by a fixed formula and depend on what the enemy does. The target moves and may require counteraction: rules of organization must be transcended.

The animal depends on an outside source for its food supply. Its freedom of motion has been gained by giving up the plant's freedom

---

* A tetrahedron, or four points, is required to establish a rigid structure. To store energy in a rigid structure, one more point is required.

to manufacture its own energy (photosynthesis). The animal not only must obtain food, but must break it down (digest it) and assimilate it into its own system. This transformation of food dictates the need for the animal's digestive, circulatory, and respiratory systems, consisting of organs whose cooperative interaction enables it to convert ingested food for its own uses. These energy-converting functions are actually the first to evolve. As we will see in studying the substages of the animal kingdom, the first animal organ to develop is the stomach. Next follows a fairly complete organ system. Legs and their musculature come later.

This dependence of the animal on an outside food supply is anticipated in process theory. It correlates to the fixity of mass of the nuclear particle. Once the animal has reached maturity, it maintains a fixed body weight, whereas the plant continues to grow.

Let us list the principal differences between plants and animals:

Plant	*Animal*
Free to create its own energy.	Committed to outside source of energy.
Committed to fixed position.	Free to move.
Size continually increases.	Size reaches a fixed value.
Shape is determined.	Shape is flexible.

Actually, 99$\frac{9}{10}$ percent of the animal's growth occurs in the womb. Growth in the womb is essentially vegetative: the embryo is rooted in the wall of the womb and grows there like a plant. The obelia (see p. 121) is invariably used by biologists to illustrate the life cycle of all animals. After fastening itself to the ocean floor, it grows in plant fashion until it releases mobile forms (jellyfish). This behavior illustrates the manner in which the animal principle builds on the cellular organization principle which is inherited from plants.

Summary

Thus the dynamic syndrome which the animal power exhibits falls under two main heads. The first is voluntary movement with the related ideas of stimulus sensing—appetite, motivation, action, satisfaction, choice, and value. These may be packaged into the single concept *animation* (from the Latin *animus*, or "spirit"). The second head comprises feeding, digestion, circulation, change of one kind of organization into another (starch into protein), metamorphosis (change from one shape to another), all of which can be packaged into the single concept *transformation*.

Both animation and transformation have a common origin, and both follow deductively from the character of level II, whose two degrees of freedom prescribe both motion and the transformation involved in digestion of food. (It is of interest that in mathematics movement is equivalent to a transformation, as in "transformation of coordinates.")

But the abstraction involved in such use of deductive principles is demanding, and it is rewarding to find that in the simplest of animals, the amoeba, the principles of motion and digestion are *fused into one* or, more correctly put, are seen as they are before the two functions differentiate.

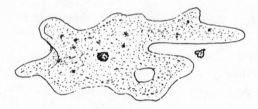

amoeba eating food particle

This is the animal power. Stimulus, motion, ingestion, and transformation are potentially present here at the first substage of the kingdom, before the multicellular creature has differentiated digestion, circulation, respiration, sensing, motion, and self-limitation into separate functions. It is as different from the vegetable principle of organization, growth, and reproduction as a principle can be, even though the animal in its cellular or vegetative organization incorporates the growth and reproductive powers of the plant kingdom.

A question of sequence: a-a-b pattern

But a disconcerting fact emerges, one that challenges the whole notion of sequential development. This is that primitive animals are very similar to primitive plants. Both are unicellular and appear to stem from the same origins.

A diagram would show molecules going through their whole sequential development to DNA, which becomes the common basis for both plants and animals.

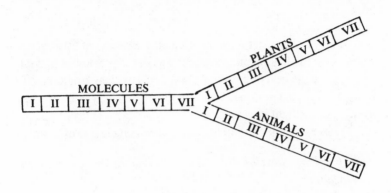

One cannot say that plants and animals form a sequence when both start from the same point, the unicellular protist.*

This objection, which apparently denies the validity of a chain of evolutionary forms, actually turns out to provide one of the most important features of process theory. This is the a-a-b pattern, which I have described in Chapter VI in regard to the periodic table of the elements.

Plants, to be sure, develop from the most complicated molecules. But if we go back further still and ask "On what sort of atoms are molecules based?" we find that 99 percent of important molecules are composed of the simplest atoms. All carbohydrates and fats are composed only of hydrogen, carbon, and oxygen, elements from the first two rows of the atomic table. If we add nitrogen and an occasional sulphur, we can construct proteins; and if we add phosphorus, the complex DNA. So that by far the greater portion of complex molecules are made of the *simplest* atoms. And complex molecules of radium atoms, or even of sixth-row atoms, scarcely if ever occur.

* Protist is the biologist's name for the hypothetical progenitor of plant and animal cells.

So there is a precedent for the branching of animals and vegetables; the molecules also branch:

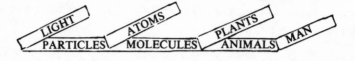

To complete the pattern, we find that all atoms except hydrogen develop from the helium nucleus at the *end* of the previous kingdom (according to the best tentative division of the nuclear kingdom so far); and we find that the proton is based on the high-energy photon at the *beginning* of the light kingdom.

So we have a regularity and hence a rule which may be formulated:

Odd kingdoms develop from the advanced end of the previous kingdom.
Even kingdoms develop from the beginning of the previous kingdom.
In other words, even kingdoms repeat, odd ones innovate.

This pattern is very explicit in the substages of the atom (covered in Chapter VI).

Helium Beryllium Magnesium Calcium Strontium Barium Radium

Helium	Beryllium	Magnesium	Calcium	Strontium	Barium	Radium
2	2 2	2 2 2	2 2 2 2	2 2 2 2 2	2 2 2 2 2 2	2 2 2 2 2 2 2
		6	6 6	6 6 6	6 6 6	6 6 6 6
				10	10 10	10 10 10
						14

Note that each type of shell repeats before a new type of shell is added. Because of this repetition, the four types of subshell, having two, six, ten, or fourteen electrons, suffice for seven rows of atoms.

But *why* does process use this a-a-b pattern for its development?

I believe the answer is that such repetition is required in order that process may incorporate as it advances. It progresses by an alternation of innovation and recapitulation. In music we encounter a similar development: a theme is repeated before new and more complex development is introduced.*

* This cumulative development is discussed in reference to art and artifacts by George Kubler in *The Shape of Time*, New Haven: Yale University Press, 1962.

Needles and pins
Needles and pins
When a man marries
His trouble begins.

It is also a device used in fairy tales, where it is always the third son
who overcomes the evil magician and marries the princess. If it be said
that this is a psychological quirk, then we have to say that nature is
psychological. Certainly, process requires memory of what it has been
through, and recapitulation achieves this.

The a-a-b pattern, then, is intrinsic to process; we will find that it holds
even in the substages, and that it provides a further theoretical constraint
on the properties of the stages, whose nature flows from these
principles.

Animal substages

First substage

The first substage of the animal kingdom, as with plants, is unicellular.
Amoebae and protozoa are examples. Both are capable of motion and
of digesting food. It is interesting how the unicellular creatures anticipate
the later development: thus, a foraminifer, despite the fact it is only one
cell, makes a spiral shell quite similar to that of snails and cephalopods
which are multicellular and billions of times heavier. The epidinium,

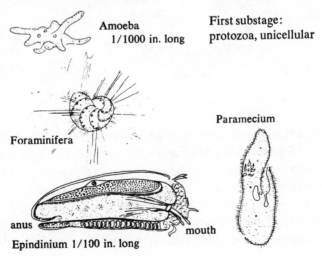

Amoeba
1/1000 in. long

First substage:
protozoa, unicellular

Foraminifera

Paramecium

anus   mouth
Epindinium 1/100 in. long

1/100 of an inch long, has a rudimentary skeleton, a mouth, stomach, and anus, anticipating the highly evolved vertebrates. Thus the first substage in its anticipation of later developments illustrates the key word "potential" of the first substage.

Second substage

Binding power produces the first multicellular animals, sponges; mobility is sacrificed in the interest of size. There is some cell differentiation, such as holdfast cells, but no organs or tissue differentiation.

It could be objected that sponges are less mobile than amoebae and hence do not represent an advance of the mobility power. But we must realize that the principle of the fall still applies to the substage: means must be developed. The motion of the unicellular animal, like its size, is microscopic. To get further, the animal must, like the plant, invest in greater bulk. The sponge also recapitulates the plant in that it is glued to the ocean floor.

Second substage: porifera (sponges), multicellular

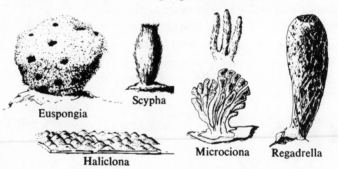

Euspongia

Scypha

Haliclona

Microciona   Regadrella

Third substage

Plants at this substage, as we have noted, develop identity in terms of reproduction. They differentiate the embryo from the rest of the plant. What does a mass of animal cells do to obtain an identity? It forms itself into a hollow chamber, or stomach, with a mouth surrounded by sensitive feelers. On the approach of a small fish or other food, the feelers seize it, draw it into the stomach, and digest it. Such is the coelenterate ("coel" = hollow, "enterate" = stomach), the first animal

Third substage: coelenterates, single organ

A Variety of Coelenterates
(adapted from Schmeil)

with an organ. The coelenterates include the sea anemone, which for a great part of its life cycle is stuck to the ocean floor and resembles a flower. (This could be evidence of a recapitulation of the previous

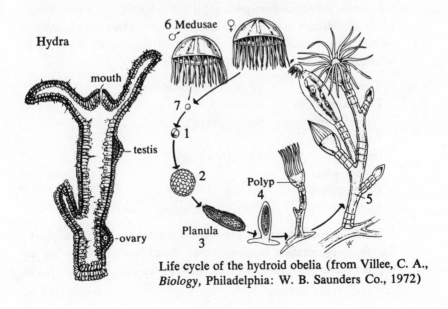

Life cycle of the hydroid obelia (from Villee, C. A., *Biology*, Philadelphia: W. B. Saunders Co., 1972)

kingdom.) Another coelenterate, the hydroid, which also looks like a plant, produces buds which break off and become mobile jellyfish that use the same hollow organ to propel themselves. Its life cycle, viewed in terms of process theory, falls into seven stages: (1) starts as one cell; (2) becomes multicellular; (3) acquires a shape (identity); (4) fastens to the ocean floor; (5) grows in plant-like fashion; (6) flowers break off into mobile jellyfish; (7) fertilization.

Fourth substage

Up to this point our subdivisions correlate with the divisions made by zoologists: (1) unicellular, (2) multicellular, and (3) coelenterates. But in our scheme the fourth substage, defined as "combination of organs," includes animals which zoologists now divide into a number of phyla.

The classification of animals by zoologists into phyla, which is the broadest division of the animal kingdom, is based on morphology (structure). The classification used in the present theory is based on the *degree of organization*. On the whole, the two classifications agree well. This is because the classification by organization, i.e., one-cell, multicell, etc., inevitably creates morphological distinctions. However, the reverse is not true; morphological differences do not necessarily imply a difference in level of organization. Thus there are a number of animal types which we list as fourth-substage (many organs), but which zoologists divide into different phyla:

Ctenophora	(Possible third-substage)
Platyhelminthes	Flatworm
Nemathelminthes	Roundworm
Rotifera ⎫	
Bryozoa ⎬	Partially extinct
Brachiopoda ⎭	
Echinodermata	Starfish
Mollusca	Shellfish (including squid)

It would be difficult to list the above animal types in order of their degree of organization. Some are an advance in one respect, others in another; none has made the organizational leap that distinguishes the

Fourth substage: combination of organs

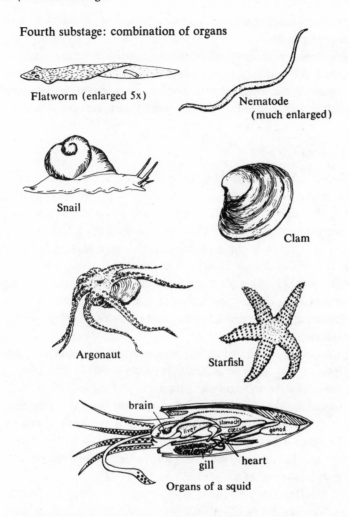

Flatworm (enlarged 5x)

Nematode
(much enlarged)

Snail

Clam

Argonaut

Starfish

brain

liver   stomach
          coecum   gonad

gill   heart

Organs of a squid

annelid or earthworm: that of arranging organs into a hierarchy, with a head at the top.

The division we make, on the basis of organization, puts all the above in the same substage—*combination of organs*. It thus emphasizes *combination*, the key word of the fourth, and it also meets the a-a-b requirement since the innovation of the third substage, the organ, is here repeated.

Admittedly, there is a great variety of creature here, but the variety *per se* does not bring a new *level* of organization, which is what the substages represent.

An interesting feature of this substage, which echoes the combination theme in a different way, is that most of these creatures are of two sexes which mate, in contrast to the third-stage coelenterates, which reproduce by budding, and to the fifth substage, which are hermaphrodite.

Fifth substage

The fourth-substage animals develop all the organs which are required; how is organization to do more? It cannot improve the organs themselves, but it can make them more effective by providing a chain. The fifth principle or power is chain-like: a chain of molecular units as in the polymers, a chain of bamboo-like segments as in the calamites (fifth-substage plants), a chain of cells or of generations throughout the plant kingdom. Here we can rejoin the zoologists, and use the classification they provide. We can place at the fifth substage the first animals which consist of a chain of segments or rings, known as *annelids*.

The annelids, which include the common earthworm, involve a new principle, known as *metamerization*, which persists in all higher kingdoms. (The vertebrates are metameric.) Metamerization involves the arrangement of organs in sequence, with the head at the top and close to the mouth to deal with new situations; what was the stomach now becomes a long passageway with organs of digestion and assimilation, having semiautomatic jobs which do not require choice or attention, located at a distance from the head.

Annelids thus meet the chain requirement by creating a hierarchy of organs. Their mobility, based on stretching and contracting, is also linear in nature.

We might also note that of the two basic contributions to evolution, transformation and mobility, the power to transform (i.e., obtaining energy by digestion of food and its conversion into new forms) evolves first, on the downward branch of the arc. The power to move, though present in all animals, begins its development at the fourth substage and continues through the upward branch. Here with the annelids it makes a great leap; by providing articulation it lays the basis for the next stage, arthropods.

Fifth substage: annelids, hierarchy (chain)

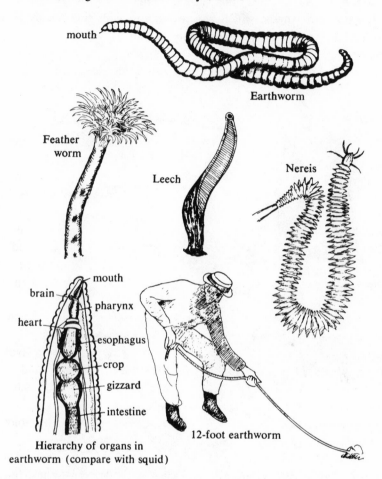

mouth

Earthworm

Feather
worm

Leech

Nereis

brain

mouth

heart

pharynx

esophagus

crop

gizzard

intestine

Hierarchy of organs in
earthworm (compare with squid)

12-foot earthworm

Sixth substage

It is impossible to do justice to the variety of life by listing it under
levels of organization, especially when we come to *arthropods*, which
include crustaceans, spiders, and some 600,000 species of insects. On
the other hand, there is something wonderful about being able to
express the key principles by which so much variety is generated. Even
more so when the very names chosen by zoologists confirm the stages
of process chosen. Arthropods, which means "jointed feet," are the first

animals with legs. What are legs? They are segmented side chains attached to a segmented trunk, an echo of the sixth-substage molecules, functional polymers, which are also chains with side chains (and remember too that they are chemicals that move).

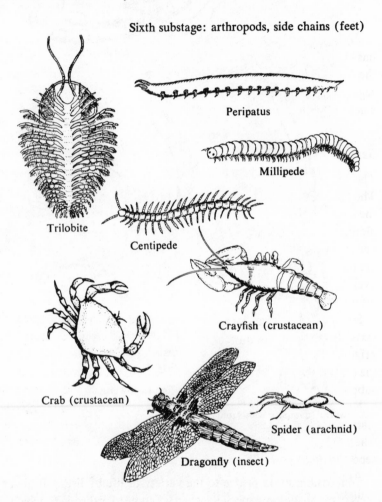

Sixth substage: arthropods, side chains (feet)

Peripatus

Millipede

Trilobite

Centipede

Crayfish (crustacean)

Crab (crustacean)

Dragonfly (insect)

Spider (arachnid)

Referring to the illustration, we can appreciate that the examples chosen can hardly convey the scope of this great phylum. But one thing is evident: legs, or segmented appendages, are the theme throughout. Such emphasis on legs must have to do with their importance for mobility. While millipedes have less than one thousand legs and centipedes less than one hundred (forty-four in the one shown), it is interesting that crustaceans (decapods) have ten legs, spiders eight, and insects six. Moreover, with the exception of the peripatus, all, including the trilobite, have legs of seven segments. (So do we.)

The examples shown are the main subdivisions or classes of the phylum Arthropoda (Decapoda are represented by two examples).

Looking back over the animal kingdom, note how the a-a-b pattern has held. The first creatures were unicellular, the second multicellular; the third had one organ, the fourth many organs; the fifth introduced segmentation, the sixth added segmented legs. Each odd stage introduces a new development; each even stage repeats and extends it.

Seventh substage

The seventh starts over with a new principle, embodied in the Chordata. The new principle, suggested by the name of this phylum, the last of the animal kingdom, is a central nervous system (notochord) extending along the animal's back and (except for certain primitive marine forms) encased in a protective bony spine. Other important distinctive features are present, but rather than recite these I would ask the reader to look over the classes of this phylum, from lancelets to mammals, and make his own observations.

So far as a knowledge of zoology goes, I can only repeat what the textbooks say, which is interesting and important. However, since the origin of the chordates is unknown, it might be just as well to look at the creatures rather than the textbook. They include primitive spineless subphyla (of which only the lancelet is shown) and all the vertebrates.

Note that although arthropods are mostly terrestrial, the chordates start in the sea. They begin over again with regard to legs too, and when they do sprout legs, the number is four, a number not in the repertoire of the arthropods despite their preoccupation with legs.

Another thing that interests me is the eye.* All have eyes, not the compound eyes of the insect, but a real eyeball. You couldn't look a bee in the eye, because he has a thousand eyes. I had not noticed the

---

* The retina and the optic nerve of the chordate eye actually grow out from the forebrain. A pair of bulbs or stalks develops and expands, each to form a hollow called the optic cup. These stalks project until they contact the outer layer of the embryo. The lens of the eye is formed from this outer layer.

Seventh substage: chordates

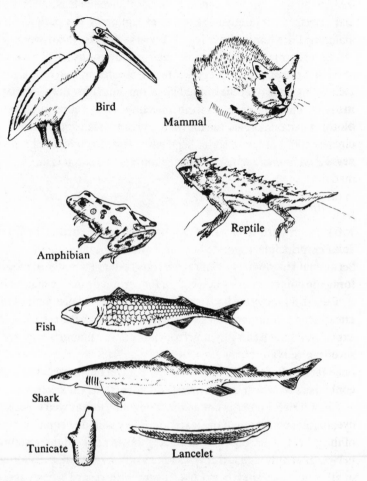

eye until I made the sketches, but once I had done so, the eye struck me as a motif as important as is the leg of the arthropods in the previous substage.

The control function

A conspicuous feature of the chordates' nervous system is that it is dual: there is the so-called *autonomic system*, itself divided into sympathetic and parasympathetic, and the *voluntary system*. These systems have different functions: the former controls blood circulation, bowel movement, and other autonomic functions, the latter the sensory and motor functions which have to do with volitional movement. The two functions were formerly regarded as separate, and it was assumed that the voluntary could not take over control of circulation, heartbeat, etc. But recent investigation has shown that, with the aid of instrumentation which tells the subject when, for example, there is more blood in the hand, it is possible to learn to increase or decrease the circulation to that member. These experiments show that the voluntary nervous system can take over the autonomic functions if an effort is made to that end. It has been found possible to train mammals (rats) in this manner. Training in such control has long been a part of Yoga techniques, which I will not go into here. What is important is that without such an extra function of control, it is difficult to explain the voluntary control of autonomous functions. The occurrence of conflict between purpose (or "spirit") and appetite could not be accounted for otherwise.

This dualism answers a number of questions that otherwise are unexplained. More significantly, it is in keeping with what would be expected on the basis of process theory. The seventh substage, like the seventh stage, must introduce a new principle. The principle must be one that has a control function. And it is appropriate that it embrace control of the organism itself, as well as control of the environment.

The voluntary nervous system can, in fact, be thought of as an inner eye, an agency of self-consciousness which cannot be explained in terms of the purely sensory awareness which would arise from the autonomic nervous system alone. As noted earlier, the eye is a conspicuous feature of all vertebrates, and its morphological and functional basis as the upper end of the nerve cord, and hence its link with the brain, adds a convincing argument that the nerve cord has its origins very early in the evolutionary chain and is not derived from annelid, and certainly not from arthropod, ancestors.

Recapitulation

The exuberance and variety of the animal kingdom may have distracted the reader from the main reason for the survey, so we should recall that our purpose was to illustrate grid technique, which is based on the assumption that each stage of process is itself a process and that the character of each substage is similar to the corresponding stage, and vice versa.

We have seen how the animal evolves in stages or levels of organization. It begins as a single microscopic cell, very much alive but with a very small radius of action. Its next step is to gain size by becoming multicellular. Size attained, it acquires identity through creation of a stomach, then other, coordinating organs: heart, liver, intestine, etc. Its next step is metamerization: the body laid out in a chain of segments, as in the earthworm, with the organs in sequence, like an assembly line. It is now ready for the innovation that makes true mobility possible, the articulated foot (arthropod). (Certainly if man's greatest invention was the wheel, nature's was the foot!) Finally, the whole organism is coordinated under the central nervous system of the Chordata. Note that it is just this kind of evolution, the advance to more complex form, that is not explained by survival. The leaps to higher forms of organization are inherent in all process, atoms and molecules as well as living creatures.

X | Protoplasm and psychic pseudopods

Having drawn on the evidence from science concerning areas in which phenomena are subject to laws, we should now venture ahead into areas where science has not found laws—or where the only known laws are of a statistical nature.

Process theory and the laws of the four levels

It is generally assumed that all phenomena are subject to law, and that the inability to find such laws is temporary and will be conquered. This presumption still survives in scientific opinion, despite the fact that quantum theory has shown that there are areas in which laws do not hold and cannot in theory apply.

But as was emphasized in Chapter IV, the "fall" into manifestation, by which process gains means for its fulfillment, is necessarily a fall from an initial *freedom*. This freedom, while random and ineffectual, is the basic ingredient, and law is the *means* by which process attains the competence to obtain its goals. If we can bring ourselves to think of the determinism of the molecular kingdom as the stepping stone to higher levels of organization and to the recovery of freedom, we can expand our science to cover what would otherwise remain inexplicable. Our theory of process therefore differs from the scientific usage in stressing the importance of freedom as well as of determinism.

Another way in which our method differs from that of science is that we are dealing with *individual* entities. When we refer to the "freedom" of photons or of fundamental particles, we mean *single* photons, which, it is correct to say, are unpredictable; or single electrons, which are unpredictable in their position. This would not be questioned by science; in fact, it is a major finding of science. But science, in its emphasis on law, takes refuge in statistical laws or probabilities, much as do insurance companies which predict the percentage of people who will die before eighty, but not *which* persons will die.

The contribution of the theory of process is that it provides four categorically distinct realms in which law applies in different degrees. These we call the four levels. Level I, with the exception that its photons differ in energy, is free of law. Levels II and III are partially free and partially determined (as described on p. 34), level II having two degrees of freedom (position) and level III one degree of freedom (energy). Level IV is completely determined by law. So characterized, the levels describe the principal features of the kingdoms (see Chapter IV).

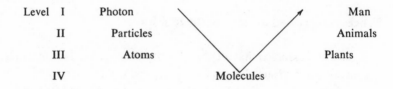

Level I	Photon		Man
II	Particles		Animals
III	Atoms		Plants
IV		Molecules	

That is to say, both the growth of plants and the radiation or absorption of energy by atoms are characterized as one degree of freedom, and both the uncertainty of position of proton and electron and that of animals are characterized by two degrees of freedom.

But we are now ready to see these levels in a way that applies more directly to the animal and the human situation. Thus, level I is purpose, level II is motivation, level III is concept formation or intellect, and level IV the physical body, especially in its ability to react to other bodies and thus provide feedback. Of these four levels, the last two (III, IV), are objective: concept formation in the sense that it can be communicated, and the physical body in the sense that it is an object.

It is with the first two that difficulties arise, since both are nonobjective: the photon (or purpose) because observation annihilates it (if the spy's purpose is known, it is annulled), and the nuclear particle because it cannot be identified. Is the electron that came out of the atom the same as the one that went in? Is the dollar I withdrew from the bank the same as the one deposited? Rather than refer to such nonobjectivity as subjective, which would imply that it is interior to the person, I prefer to call it *projective*, which means only that it is not objective. As we have often stressed, the universe contains this projectiveness. Thus a machine or a vehicle has a projectiveness in its purpose. It must have motivation (the fuel). Similarly, a bullet is aimed (purpose) and propelled by the charge (see p. 7 and also the beginning of Chapter IV).

	Human	*Vehicle*	*Bullet*	
Level I	Purpose	Control	Aim	} Projective
Level II	Emotion	Fuel	Charge	
Level III	Intellect	Plan	Computation	} Objective
Level IV	Body	Hardware	Bullet	

Our problem is to expand science to include the two projective factors, purpose and motivation, in order to better understand *principles* which contribute to cosmology and are essential to life.

The phenomenon of motivation

In this chapter we will discuss phenomena which have to do with the second of these factors, *motivation*, as it occurs in animals and man. These phenomena are of particular interest because of their nonobjective component, which we should no longer have to reject categorically, for we now expect something of a projective nature to exist.

Let us list some of these phenomena:

1. The motion of an amoeba.
2. The emotional projections of persons.
3. The behavioral pattern of animals (instinct).
4. Certain aspects of extrasensory perception.
5. The dream state.

6. The use of sacrifice in primitive religious ceremonies.
7. Psychic healing (by laying on of hands, or at a distance).
8. The ectoplasm of materializing mediums.

I have listed the phenomena in order of the credibility of my use of them from the viewpoint of current rational thinking. It hardly needs saying that some of them are not considered factual, and because they are not, no theory is developed for their explanation. Nor do I ask the reader to credit them all; I ask only that he consider them together.

As to proof, we are in the predicament of having no court of appeal. To take but a single case, the facts of physical materialization by mediums were repeatedly and dramatically confirmed by reputable scientists. Among these is Gustav Geley,* who in 1924 produced casts of ectoplasmic hands interlocked in such a manner that it would be impossible to duplicate by known means.

In order to prove the existence of materializations, Geley made wax molds of hands materialized by the medium. When the wax had hardened, the hands were dematerialized. Geley then poured plaster into the wax mold and, when the plaster had set, melted the wax. (The reverse of the lost-wax process.) He had the testimony of experts that such molds (of folded hands) were such that the hands could not be *withdrawn* from the mold and were therefore dematerialized (one glance at the photographs confirms this statement).

All necessary precautions against fraud were taken, and some of Geley's experiments were witnessed and testified to by a panel of thirty-four scientists and officials.

More empirical proof can hardly be imagined, yet this work has been totally ignored. Why? Because *there is no theory to account for it*, and existing theories apparently rule out its reality.

This is but one of many examples, and what is common to each of these examples is that they cannot be explained by modern science.

Schrenk von Notzing also worked with materializing mediums and even captured ectoplasm and examined it under the microscope.** My purpose is not to say that these phenomena have been proved, but to

---

* Geley, Gustav. *Clairvoyance and Materialization*. London: Allen & Unwin, 1927.
** Schrenk von Notzing, A. P. F. Baron. *Phenomena of Materialization*. Translated by E. E. Fournier d'Albe. London: Kegan Paul; New York: Dutton, 1920.

point out that however much proof was or could be provided, the phenomena are not acknowledged as facts because they are so drastically at variance with the prevalent interpretation of science.

The criterion of falsifiability

Philosophers of science are given to pontificating at length on what they call the criterion of falsifiability: the thesis that for a theory to be a scientific theory, it must be possible to subject it to test (that might prove it false). It does not occur to them that this weapon might actually be used against the basic postulates of science. Instead, they tend to brand all ESP phenomena as fraudulent on the grounds of being contrary to these postulates of science.* One is reminded of the medieval doctors who denied Galileo's telescopic observations.

This attitude is against the interests of true science and is even contrary to elementary justice, for it becomes impossible to correct a theory by experimental test as long as theory decrees in advance what the outcome of the test must be.

The motion of the amoeba

But certain of our listed items to which the hypothesis of an animating and directed energy applies *are* recognized as facts. One is the amazing instincts of animals (some moths navigate by the stars!). Another is the motion of the amoeba. The latter is one of the simplest facts of animal behavior, yet the amoeba moves without muscles! Actually, motion with muscles in more highly evolved animals only conceals the mystery. What activates the muscles? Nerves, of course, but what activates the nerves?

The following is a typical account of amoeboid movement:

> Locomotion is accomplished by forming temporary pseudopodia and is known as **ameboid** movement. At the beginning of a pseudopodium is a fingerlike projection of ectoplasm (*sic*); then the granular plasmasol flows into the projection as it lengthens. Mast's theory, based upon the colloidal

---

* As a matter of fact, I find it difficult to discover what law of science denies the phenomena of telepathy, precognition, etc. It is, rather, with alleged implications of the laws of science that such phenomena conflict.

nature of protoplasm, states that the movement is based upon the reversible change from the fluid sol state to the gel state. In this process the posterior part of the moving Amoeba is changed from the plasmagel to the plasmasol, while just the reverse is occurring at the anterior end where the pseudopodium is forming.*

This explanation does not tell us what changes the plasmagel to plasmasol, nor how the amoeba can exhibit choice in selecting food. No matter. What happens is observable. The amoeba, described as a mass of clear colorless jelly, forms temporary extensions called pseudopods at any place in the cell body.

Psychic protoplasm

Raymond Prince, in a recent article,** deals with attention and likens psychic energy to protoplasm which is activated and directed by attention. He quotes from Freud, who made a similar analogy between one-celled creatures and libido:

> Freud . . . was fond of comparing the ego to an amoeba whose body was in a libidinous reservoir stretching out and withdrawing its libidinous pseudopods as interest was focussed and relinquished. In his words:
> "Think of the simplest forms of life, consisting of a little mass of only slightly differentiated protoplasmic substances. They extend protrusions which are called pseudopodia into which the protoplasm overflows. They can, however, again withdraw their extensions of themselves and reform themselves into a mass. We compare this extending of protrusion to the radiation of libido on to the objects, while the greatest volume of the libido may yet remain within the ego: we infer that under normal conditions ego–libido can transform itself into object–libido without difficulty and that this can subsequently be absorbed into the ego."***

Let us return again to Geley.

Geley, in 1924 in the introduction to his work, *Clairvoyance and Materialization*, rejoices in the then current shift to experimentation

---

* Beaver, W. C., and Noldand, G. B. *General Biology* (p. 237). 7th ed. St. Louis: Mosby, 1966.

** Prince, Raymond. "Interest Disorders." *Journal for the Study of Consciousness*, vol. 4, no. 1 (Spring, 1971).

*** Freud, S. *A General Introduction to Psychoanalysis*. New York: Doubleday, 1920.

with mediums and away from "mystical theories." In the reports of the
Metapsychic Congress of Copenhagen (1921) and Warsaw (1923), he
points out "there is no allusion to phantasms of the living and the dead,
to spirits, etc. All speak very simply of a biological phenomenon," to
which "there appear to be analogies, or at least points of contact,
between the ectoplasmic process on the one hand, and normal
physiology, animal biology, and certain phenomena classed among the
natural sciences, on the other." And he continues:

> "Materialisation" is therefore no longer the marvellous and quasi-
> miraculous affair described and commented on in early spiritist works; and
> for this reason it seems to me desirable to substitute for "materialisation"
> the term "ectoplasmic form." . . .
>
> What is an ectoplasmic form? To begin with, it is a physical duplication
> of the medium.
>
> During a trance a portion of his organism is externalized. This portion
> is sometimes very small, sometimes very considerable—amounting to half
> the weight of the body in some of Crawford's experiments. Observation
> shows this ectoplasm as an amorphous substance which may be either
> solid or vaporous. Then, usually very soon, the formless substance
> becomes organic, it condenses, and forms appear, which, when the process
> is complete, have all the anatomical and physiological characters of
> biologic life. The ectoplasm has become a living being or a fractional part
> of a living being, but is always closely connected to the body of the
> medium, of which it is a kind of prolongation, and into which it is
> absorbed at the end of the experiment. Such is the bare fact considered
> in itself and apart from certain complications which will all be studied
> later. It is the naked fact, dissected, so to speak, down to its anatomical
> and physiological structure. This fact is substantiated, with formal proofs,
> by the common consent of scientists from all countries.
>
> The objective reality of these forms is proved by photographs taken by
> flashlight, by their imprints on clay, on lamp-black, and on plaster, and
> finally, in some most notable cases, by complete casts.
>
> The phenomenon is the same in all countries, whoever the observer or
> the medium may be. Crookes, Gibier, Sir Oliver Lodge, Professor Richet,
> Ochorowicz, Professor Moreselli, Dr. Imoda, Mme. Bisson, Dr. von
> Schrenk von Notzing, Crawford, Lebiedzinski, myself, and others, all
> describe exactly the same thing.*

Even this long quote does not do justice to this remarkable book, nor
of course to the subject. Schrenk von Notzing describes similar findings
in his equally important work.

* *Ibid.*, pp. 175–176.

In short, then, there is confirmation for the existence of an "ectoplasm." The question of what it is I cannot answer. What I am trying to point out is that there is a theoretical *requirement*, as borne out in Freud's selection of the protoplasmic movement of the amoeba to illustrate the behavior of the psyche, and there is also *evidence* (as in the phenomenon of materialization) for the existence of a formless and animate substance analogous to the undifferentiated protoplasm of the amoeba.

Such a substance is just about what is to be anticipated deductively in the sixth principle of process. It will be recalled that one of the key words used to indicate the essence of the second principle is *substance*, and that it has to do with attraction and repulsion, as in electrical charges, but that at the second stage it is not under control. Its controlled version should, and does, occur at the sixth stage where, according to our theory, it supplies the animated principle of animal life.

Permit me now to adduce another quote, this time from an earlier draft of the present book, written at a time when I did not know about Freud's choice of the analogy of the amoeba to the psyche and when its similarity to the ectoplasm of materialization phenomena had not occurred to me.

> We can now discern that, so defined, the "animal body" is not necessarily or even partially a physical object as is the physical body, which, of course, belongs to stages four and five. We call hunger a physical urge and, compared with mind, designate it as more material, more tangible. But compared with a stone, or even with the physical body itself—a "true" physical object—the "animal body" is not objective. It is obviously *dynamic*, a syndrome of urges, pulls, forces, etc. In the use of the term "field," whose scientific and therefore "objective" implications were discussed in the preceding chapter, we are now in a position to supplement the objectivity with direct testimony. . . . As applied to the animal we could use "field" to describe a sphere of influence or interaction extending beyond the body of the animal that determines the animal's orbit of interest. This is a legitimate tentative use of the word.
>
> Working up from the molecular level we find that at the molecular level there was coincidence of the organizing* principle and the embodiment.

* In this particular passage I used the word "organizing" to cover molecular combination, plant growth, and animal mobility.

At the vegetable level the organizing principle extended sufficiently beyond the physical border of the cells to organize multicellular growth (we do not know how, it might be electrical [Burr, etc.], but the problem we are posing now is not to explain it so much as to recognize that it exists). When we come to the animal kingdom, we must recognize that this organizing principle extends still further, enough to provide a basis for interaction with environment and hence for development of senses and the reach-and-withdraw animal activity. (What I now call animation.)

We arrive at this notion of an organizing field that extends beyond the boundary of the physical body by reasoning about what we know of vegetables and animals. We cannot see or touch this organizing principle. The evidence for it is the creature itself. Like the evidence for a magnetic field, which cannot be seen or touched, the presence of a field is detected by introduction of special materials that the field responds to. Thus we detect a magnet with iron filings or a compass.

This should suffice to show the similarities of amoeboid movement, "psychic protoplasm," and ectoplasm to the postulated sixth-stage "substance," or animating principle, which I was endeavoring to describe before having read Freud or Prince.

Normally, one expects things which exist to be visible, but science has extended existence to include entities like force, charge, energy, which are not directly visible. But the fact that ectoplasm, which we would not expect to be visible, can under certain circumstances be photographed is important to our theory because it emphasizes the physical or substantial nature of the animating principle.

The possibility that ectoplasm can be photographed, although not adding to the credibility of materialization, should not be set aside. The quasi-physical nature of ectoplasm may provide the important clue to understanding the unsolved problem of mind–body interconnection. We should withhold judgment until later, when we know more about it either factually or theoretically.

The need for a medium

Let us return to our list of phenomena which have to do with motivation. Two of the items, the use of sacrifice in religious ritual and psychic healing, provide clues. It will be recalled that the protoplasm in the amoeba extends pseudopods at any point in the body of the amoeba and in any direction toward food particles and the like. Mast's explanation may be a correct account of the mechanism of the movement, but it does not dispel the indication that behind these changes there must be the act of will, attention, or intention, which initiates and directs the pseudopods. A similar act occurs within the psyche, as it does in materialization. Thus the ectoplasm is an *agency of intention*. It is the passive element through which attention or intention manifests.

I owe to Dr. Oscar Brunler the only explanation I have heard of the reason for animal sacrifice. He held that at the death of the animal, a psychic substance was released which made it possible to communicate with the dead—in the case of the religious ritual, with the ancestors or former leaders of that civilization or tribe. Dr. Brunler, to substantiate his view, called attention to the sexual orgies which follow when there has been a mass execution or when many are killed in battle. The "ectoplasm" released on such occasions becomes available and accentuates sexual energies.

A similar use of ectoplasm, Dr. Brunler maintained, occurs with mediums, who make their own ectoplasm temporarily available to the "spirits" for the purpose of communication.

Even if we withhold judgment as to the validity of this explanation, let us at least note its consistency with the overall theory we have constructed, for in the cases cited, as in others, there is a *gap* between the directing principle and the physical world. The directing principle requires in all cases a *medium* for communication or other intercession with the physical universe.

The fact that under normal conditions a living creature activates its own organism and interacts with the environment without visible evidence of any medium being involved does not mean that a medium does not exist. Its existence, in fact, fills just the gap that Freud was describing by his analogy of the psyche to the amoeba.

Dreams

This brings us to another item on our list which we have not yet discussed, dreams, which I believe furnish additional evidence, but this time from the subjective point of view.

In dreams we are cut off from the outer world of the physical senses. It is significant that we are not able to run in a dream when we want to, and that if we do "make our muscles do what we want," we immediately wake up. From this it is clear that the part of the brain that governs motor action and control is minimally functional during the dream state. The act of trying to run causes the cerebellum to be switched on and wakes us up. It also seems evident, from the fact that dreams are often completely devoid of correct associations, that the cerebral cortex, which governs learned association, is not functioning. Durham (whose findings are quoted in the eleventh edition of the *Encyclopaedia Britannica*) made measurements of blood circulation in the brain during sleep and found that circulation dropped almost to zero. More recent work involving the encephalograph and measures of electrical activity of the brain during sleep has been interpreted as indicative of brain activity, but if so it is quite different from waking activity.

Yet in dreams we do have vivid experiences, visual, emotional, even aural. We move, fly, have adventures, anxieties, sexual encounters, ordeals, agonies and ecstasies perhaps even more interesting and varied than those in normal life; and yet these experiences are not due to, and are apparently *without even a contribution by, the brain*. Indeed, it would appear that the brain functions as a load does on a piece of machinery. It ties it down to a specific function. Take the load off a motor and it "races." Or perhaps we should say the brain regulates and channels the psychic activity which in the dream roams about, like a horse out of harness. Psyche and brain are separate.

Emotional projections

Similar indications arise in the so-called sensory deprivation experiments in which the person being tested is kept in a dark, soundless environment, sometimes suspended in warm water so that he can have no feedback to give a sense of muscular orientation. A few days of this kind of thing produces a raving maniac. The person is invaded with all kinds of wild visions not unlike a "bad trip" with LSD.

This is more evidence to show that we are bathed in a world of imagery that is held in check only by the waking mind.

We say that these phantasms arise in the unconscious, but where and what is the unconscious? What is the "substance" from which these vivid hallucinations are formed? To call them unreal or to say that they arise in the unconscious does not explain them.

And what do we know of how prevalent just such hallucinatory imaginings may be in ordinary daily life? I see across the street the back of a fascinating female creature, quicken my steps to catch a closer glimpse; then as I get closer, or see her turn to look in a store window, I see that my imagination has played me false. The girl is quite homely.

Or, again, we are told by psychologists that the new-born chick does not really have a true perception of its mother. It will follow any object of appropriate size, such as an automated football. This deduction may satisfy the psychologists' instinct for mechanical explanations, but for me it rather suggests that "mother" is a subjective idea, an archetype of the chick world, and that this archetype exists prior to the training of sense experience which will eventually make its contribution, but subsequent to what is subjective, or archetypal.

Such considerations fall into place as giving us an inside view of the activities of this psychoplasm which we may also call imagination. Operationally, it is an animating principle; subjectively, it is populated with archetypal, or "typical," images, which attach themselves to or become associated with external objects and thus motivate their possessor toward or away from these objects.

Nucleation through attraction

But there still remains an important point. Let me quote again from an earlier draft of this present book:

> The animal moves—or rather, let us say, chases rabbits. Now it is entirely pertinent to say here that people spend money to see greyhound races, and that the greyhounds are induced to race by a mechanical rabbit that runs around a track in front of them. The fact that the rabbit isn't real makes no difference to the spectators or to the dogs. As long as the dog believes the object bounding in front of him is a rabbit, he runs. Wherein for the dog is the substance? "Faith is the substance." . . .
>
> Here at last we have a clue: the dynamic which motivates the dog "nucleates" as a rabbit, a chase-inducing item having an attractiveness just as a positively charged nucleus has for the electron. Perhaps the Playboy "bunnies" are another example.
>
> In other words, we cannot think of the animal power as only a *dynamic*. It must nucleate as something. It involves a triplicity of object, act, and anticipation, analogous to our second-kingdom triplicity of mass, motion, and charge. In other words, substance is as much a part of the picture as energy is. In fact, *substance is condensed psychic energy*.

The notion of nucleation is helpful in the tendency of the plastic substance to "fixate" or jell into an image. This "fixing" for the second stage is the particle itself. For the sixth stage, it is the target animal, the mate, the quarry, or whatever object or image becomes endowed with a charge.

> Such tricks hath strong imagination,
> That, if it would but apprehend some joy,
> It comprehends some bringer of that joy;
> Or in the night, imagining some fear,
> How easy is a bush supposed a bear!
>
> *A Midsummer Night's Dream*, Act V, scene i

For man, or for human consciousness, these nucleations are what Jung calls archetypes. They are carriers of a charge, endowed with either negative or positive value, which in the psychologists' language produce a stimulus or induce reactive patterns, or drives. Seen in terms of man's larger development, they are condensations of emotional energy whose compulsive power can, however, be overcome and their captive energy released by calling them into consciousness and understanding their nature. Most psychotherapy has to do with unlocking these fixed "nucleations" of energy, which compare to free psychic energy as mass to kinetic energy.

The reader will note that my use of the word *nucleation* is intended to show the correspondence of stage six to stage two. These two stages, both at the second level, have to do with condensation of energy, and I am correlating the condensation of energy into nuclear particles with the condensation of psychic energy into archetypes.

Now, as Raymond Prince points out (in a passage immediately following his quote of Freud's comparison of the amoeba's pseudopods with human libido radiations), Freud used a term that makes the parallel to nuclear particles even closer than I have made it, although he had no such parallel in mind. Prince writes:

> Within psychoanalysis there has been a good deal of discussion about the fundamental nature of the "psychic energy," the "protoplasm" employed in the various activities of the ego. . . . In speaking of this general area, Freud employs the German *besetzung*, meaning "to invest with a charge."*

* Prince, Raymond. "Interest Disorders." *Journal for the Study of Consciousness*, vol. 4, no. 1 (Spring 1971).

Thus Freud, endeavoring to describe energy, draws on the concept of *charge*, which is the third of three important linked concepts attached to stage two (motion, mass, and charge). The word *besetzung*, meaning *to invest with charge*, conveys just what differentiates the sixth stage, which we call volitional, from the compulsive second stage. The sixth uses charge, whereas the second is used by it since the electron is attached to the proton compulsively. Even with electricity there is the phenomenon of "induced charge," where the presence of a charge "induces" the opposite charge in a neutral body.

Freud supplies the point I missed in the example above about greyhounds chasing rabbits, a point which I should have noticed if I had made full use of the correlation with nuclear particles; for the "nucleation" of matter at stage two is always accompanied by "charge," that is to say, by the force of attraction or "attractiveness."

The difference between attraction and attractiveness is important because it serves to emphasize the element of participation that is excluded from the cosmology of science.

A response to science

In closing, we may note that the subject we have just covered—the demonstration of the existence of a plasmic directable energy—is especially difficult because such existence is nonobjective. Rather than defend it here (we have stated repeatedly the case for the nonobjective), we can note that "what's wrong" with science is not so much that it is materialistic and mechanistic, but that it tries to be entirely objective. As William James put it:

> Compared with the world of living individualized feelings the world of generalized objects which the intellect contemplates is without solidity or life. As in stereoscopic or kinetoscopic pictures seen outside the instrument, the third dimension, the movement, the vital element are not there. We get a beautiful picture of an express train supposed to be moving but where in the picture, as I have heard a friend say, is the energy or the fifty miles an hour?*

* James, William. *Varieties of Religious Experience*. New York: Longmans Green & Co., 1916; Random House (Modern Library), 1961; Macmillan Co. (Collier Books), 1961.

To return to pseudopods, there is a most interesting and recent contribution to this subject described in *A Separate Reality* by Carlos Castaneda.* The book is one of several in which Castaneda describes his initiation into magical practices by the American Indian Don Juan.

While the whole book deals most provocatively with the magical world of psychic projection, the reference which pertains to the present chapter is the story about Don Genero crossing the waterfall. To do so, Don Genero puts out long ectoplasmic arms which he uses to support his weight and permit an otherwise impossible feat. These arms are "seen" by Castaneda. We can also recognize that the fearsome animals that Castaneda is required to confront and convert to allies are equivalent to the archetypes of Jung—they too are nucleations of psychic substance which convert to free energy where properly handled.

Of course, I cannot testify to the validity of this report, or for that matter to that by Geley and others who have experimented with mediums and testified to the existence of ectoplasmic extrusions. In fact, I've not myself seen the pseudopods of the amoeba.

I have, however, for a long time felt that in order for the sensing of animals to evolve, there must be some primordial physical extension of the self that can reach out and touch, even if only in the limited sense that a magnetic field reaches out and affects iron filings.

On one occasion—it was in a drug store—when I was in an especially "high" state, and also hungry, I directed a whammy at the back of the neck of a waitress some ten feet away. To my surprise, she cried out and put her hand on her neck, turned and looked at me. "Did you do that!" she said.

The denial by science of such phenomena does not bother me as much as the fact that even the accepted facts about animals, which pose a requirement for phenomena of this sort, are passed over in silence.

While we are not attempting in this book to deal with ESP or related phenomena, it is worth noting before closing this chapter that our revised view of cosmology with its four levels provides a foundation on which to base theories of ESP and other psychic phenomena. We have just seen how ectoplasm falls into place as having a second-level nature.

* Castaneda, Carlos. *A Separate Reality.* New York: Simon & Schuster, 1971.

The same could be said for telepathy, clairvoyance, and map dowsing. These phenomena do not show the usual dependence on distance; they behave as though there were no intervening space between the percipient and the target. This is what we would expect of the second level—where space does not exist (recall that it is not possible to attribute a precise position to a nuclear particle). Again, precognition could be assigned to level I—because here there is no time. Above all, level I establishes a basis for *intention*, important not only to parapsychology, but in life situations in general.

XI | Animal instinct and the group soul

Adjusting our sights

Having earlier considered what science can explain, and (in the last chapter) something of what it cannot, we should now have freed ourselves from the compulsion to expect all phenomena to be subject to laws of science and should be able to examine impartially evidence that is currently assumed to be explicable "any day now," but actually cannot be explained by existing scientific theories.

When science was first winning its laurels, it was in the position of having to persuade skeptics that it could legitimately be extended into areas where it had not previously ventured. The synthesis of urea, vaccination, sterilization, the chemical analysis of insulin and other proteins, and the analysis of the double helix in DNA were mileposts in this extension into what formerly was territory regarded as beyond the scope of scientific inquiry.

Presently, however, things are turned the other way about. The spectacular achievements of science have so impressed layman and scientist alike that it is assumed there is no realm which science cannot conquer. In economics, sociology, psychology, advances commensurate with this belief in the omnipotence of science have been conspicuously lacking, but no amount of failure seems to discourage the implicit faith that any day now we will have the answers.

But the theory of process gives us, to the contrary, theoretical reasons to expect that there are categorically different areas or realms between which the prevailing laws differ in the degree of predictability they afford (see p. 131). These areas or realms are the levels set forth in our arc, and we have derived their existence by close attention to the actual findings of science, especially in quantum physics. Understanding of these levels with their differing kinds of law can liberate the life sciences from their present entanglement in the billiard ball paradigm.

It could be objected that we attribute too great a significance to the uncertainty of fundamental particles, which, after all, affects only single electrons too minute to be of any consequence to living things.

We hope the reader can by now answer this objection for himself, or at least accept the relevance of the answer we would give: that the same freedom which in electrons and atoms dissipates in randomness is, on the right side of the arc, organized into what we recognize as living creatures, each stage encompassing greater scope:

Level I	Light		?
Level II	Particles		Animals
Level III	Atoms		Plants
Level IV		Molecules	

The actuality of this progression, culminating in the ascendancy of organization over the material organized, can be seen with special clarity in our survey of the molecular kingdom, where the seven substages have been worked out by science in great completeness. It will be recalled that beginning with metals (molecules with but one atom), followed by salts (in which the ionic bond operates to hold two or more atoms together in the molecule), the development then follows a sequence in which more and more atoms are held together, until with polymers there are hundreds of thousands of atoms, and with DNA millions of atoms, organized into a single molecule. Wonderful as this is, it points to what is still more so, the sweep of large-scale process from photons to man and beyond.

Thus the most thoroughly understood of the kingdoms supports the thesis of an evolution in which law is transcended and *used*, in which there is an orderly progression from entities that are under the rule of law to those that *use* law and hence are above it.

How will controls matter

But we still must find how this mastery of law evolves. For such a task, science must adjust its sights, firstly to recognize action as more basic than inert objects, and secondly to recognize the chain of causation which connects pure action, will, or purpose with physical objects. (See *Additional information* section at the end of Chapter V.)

Purpose, of course, requires means to carry it out. In the detection of ore by physical instruments, for example, we need, say, a pointer reading on an instrument, indicating some change in the magnetic field. Thus, besides purpose, the project involves physical *objects* which take a *form* to give a reading by the operation of a force. These are the three levels below level I (which is purpose itself).

Level I   Purpose
     II   Force (motive power)
          III   Form
               IV   Objects

The amoeba can be thought of similarly: the food, its *object*; the shape it makes, the plan or *form*; and the plasma moving under the direction of intention, the *force* that makes possible the change of shape. We have no trouble understanding the physical objects, nor the form or plan, nor, I trust, the self-evident nature of purpose as first cause. What eludes explanation is the nature of the moving force.

In the last chapter we pointed out the similarity of the pseudopods of the amoeba to the ectoplasmic appendages of mediums, and the psychic libido described by Freud. Now the interesting thing is that this force is no better "explained" in cases we consider quite normal, like the amoeba, than in the case of the medium. Take, for example, a person typing a letter: there is purpose, to convey a message; there are objects, the letters of the alphabet (or the typewriter keys); there is the mentally formed message. What pushes the keys? You can say muscles, or for an electric typewriter electricity, but this does not explain the cause of movement. Somewhere there is another typist (in the brain?) pushing the right nerves to move the right muscles.

Here we must avoid the trap of saying that muscular motions are conditioned reflexes and that the message is a composite of such reflexes. It is true that a great many habit patterns have been learned. The spelling of words, for example, was learned in school until it became automatic, but this automatic response becomes an agency of the voluntary self, which has relegated it to habit because it no longer requires active attention. Thousands of spellings have been so learned, along with muscular training (for writing, reading, and speaking), but above and beyond the habitual automatic reflexes there sits the voluntary self purposively writing a letter and making decisions about what to say.

We need not, and should not, try to explain this initiating principle beyond what we have already done in correlating it with the quantum of action. But how does the quantum of action, or voluntary self, which is nonphysical, reach out and press the physical keys of the typewriter? The problem is the same whether we think of the fingers pressing the keys of the actual typewriter or of the monad sitting at an imaginary console inside the brain and pressing buttons like the Beatles in the "Yellow Submarine."

This phenomenon of action acquiring the means of expression is foreshadowed by the amoeba.

Another look at the amoeba

According to current biological theory (mentioned on p. 135), the amoeba extends pseudopods by converting the plasmic substance with which it is filled from a gel (which is relatively solid) to a sol (which is fluid). But how?

At the temperature at which life can maintain itself, the average kinetic energy of the molecules in which the amoeba swims about is 1/25 of an electron volt (or about 1/40 of the energy in a flashlight battery). The wavelength of electromagnetic energy of this magnitude is about 1/1000 centimeter, which is about the size of the interior of the amoeba. In other words, the amoeba swims about in a bath of free energy, and has only to point this energy, which is heat, in whatever direction it wishes to create a pseudopod. This it can do by virtue of the free phase or timing dimension implicit in the quantum of action.

Bear in mind the "turn" discussed in Chapter V, where the molecule, by means of this same free choice of timing (the phase dimension), becomes able to store energy.* (See discussion of *phase dimension*, p. 49.) The molecule does so by controlling the forces impinging upon it. This stored energy is then made available to volition as expressed in growth. Here with the amoeba the control would appear to be achieved by converting the gel to a sol (or vice versa); in other words, by controlling the binding and thereby making a fluidic process produce a form which can effect a result.

As a plausible explanation of the observed behavior of the amoeba, this interpretation can at any rate lead us to the important interim hypothesis we need, provided we generalize or expand the idea to embrace much larger spheres than the interior of the amoeba. The amoeba is an illustration of how purpose can influence matter in a way that exhibits choice. The choice we refer to is animal choice, the choice among targets known through sense experience. (As we have noted earlier, a plant does not exercise choice except of the yes or no variety; it just chooses to grow or not to grow.)

Attraction, compulsive and controlled

This throws light on the sixth or animal power. Just as attraction brought about the fall into matter, so does the controlled use of the same force (attraction) provide for animal mobility. It is the *besetzung* (the endowing with attraction) that induces the animal to pursue its quarry or seek out a mate. Note how this differs from the compulsory attraction due to gravity or to electricity at stage two, in which there is no choice. The greyhound chases the rabbit by endowing the rabbit with the attractiveness needed to make pursuit worthwhile. If the greyhound is not hungry, it doesn't chase. One might say that hunger endows the rabbit with attractiveness, but in any case the attraction is not that of a mechanical force. (See discussion of animal mobility at the opening of Chapter IX, and the section *Nucleation through attraction*, p. 142.)

Earlier kingdoms present interestingly similar differences between

* If an analogy will help, we can again cite the self-winding wristwatch, which takes energy from the arm movements of the wearer and stores this energy to run the watch.

the sixth substage and the second. The ionic bond that holds second-substage molecules *together* reappears in the proteins of substage six, but this time to hold the two chemicals actin and myosin *apart*, that is, triggered for action, so that the ionic bond is used in reverse.

Instinct and animal soul

For animals, the sixth substage includes the 600,000 species of insects. Here instinct is a prominent development. Since animals are also the sixth kingdom, we can expect that instinct is especially emphasized in the sixth principle, mobility. Instinct,* "a tendency to actions which lead to the attainment of some goal natural to the species," meets this prescription: it is part of the mobility syndrome. Although it is virtually automatic, instinct is not necessarily compulsive. The nesting instincts of birds still leave choice operative; the bird has options. Only as it makes the nest does it proceed according to a predetermined plan.

It is currently taken for granted that complex instincts such as are typical of insects—certain moths, for example, that navigate by the stars, or the wasp that lays its eggs in the larvae of another particular species of wasp and simultaneously paralyzes the larvae by stinging them at a certain nerve ganglion—can be explained as encoded in the DNA. This is to me a serious abuse of scientific explanation. Instinctual behavior can be learned only by interaction with the environment. *Animal instinct cannot be explained by DNA.*

The role of DNA

What DNA can do is to provide a blueprint for manufacture. It encodes instructions for making chemicals which in turn build tissues and organs.

Animal behavior, on the other hand, which includes animal instinct,

---

* As the term "instinct" is used in examples drawn from human usage, driving the car instinctively or putting on the brakes instinctively, it means that a pattern of behavior has been learned (according to the trial-and-error process described in Chapter IV).

must have been perfected by trial and error. Such skill begins with play. All young animals play and so discover how to operate the physical vehicle. When Johnny the Monad gets his brand-new amoeba, the first thing he does with it is to play. This is not coded in the DNA. By the play he learns how to operate the console, how to make pseudopods, which becomes useful when he has to get food. The DNA would be capable of specifying "move toward food," but this instruction doesn't supplant the need for the practice provided by play.

The real problem is how to find and capture food, and this, DNA could not indicate, because the pertinent information has to come from interaction with the environment. The animal is to chase another animal; so the problem is with cybernetics (steering) rather than blueprints. Pursuit of a moving target cannot be computerized; it is learned by practice, much as one learns to walk.

Clearly, there is a categorical distinction between the *instructions for manufacture* which are set forth in the blueprints and *learning to use* the resultant product which involves practice, especially play, to provide the feedback on which control is based. The same may be said of behavior patterns based on instinct. These may be enormously complicated: hive-building, nest-building, migration, mating rituals, feeding, egg-laying. I cannot accept the assumption that instincts are built into the DNA, because they must have their origin in behavior and depend on interaction with the environment.

Consider again the wasp that buries its eggs in the larvae of another species of wasp, at the same time stinging the larvae at a nerve ganglion in order to paralyze them. Could any sort of blueprint teach this symbiotic relationship? If we had the kind of blueprint known as a map, we could use it to find treasure buried at a point shown on the map. But this is not how the right kind of larva is located by the predatory wasp. It must be located by the same target-seeking process that is involved in the pursuit of food.

The role of the group

The next point we must recognize is that *instinctual behavior is not learned by the individual animal*. This is clear from the fact that young animals which have not seen their parents and hence could not have

been instructed by them—young salmon or wasps—nevertheless follow
the instinct of their species. Neither can a given animal individually
transmit acquired behavior to its progeny genetically, because the genes
are isolated at an early age and cannot be influenced by the animal's
behavior.

How can we explain it? Let us look again at the arc:

Level	I	1 Light		7 (?)
	II	2 Particles		6 Animals
	III	3 Atoms		5 Plants
	IV		4 Molecules	

It will be recalled that level I is outside time. Level II exists in time,
but has no beginning or ending: its energy is transformable, but
indestructible; so likewise the fundamental particles. Level III covers
forms: its entities can be constructed and taken apart. Their existence,
in time as well as space, is finite. Since the sixth or animal principle is
at level II, it is not finite; like energy, it may change form, but cannot
cease to exist.

What does this mean? Surely an animal's existence is finite. It is born,
it dies. That is according to existing conceptions, but not according to
the theory we are presenting. Based on what we can deduce about the
levels, the animal *principle* (as distinguished from the cellular body of
the animal) must continue to exist after the cellular organism dies.
This implies that the animal principle is quite distinct from the cellular
organization with which we normally identify it.

Animal death is abrupt and distinct. The cellular organism may still
be there for a time afterwards and may even remain alive in the sense
that cells continue to grow (hair and fingernails continue to grow after
death); but from the moment after death the *animal* (animating
principle) is gone. The death of a plant is quite different. It is impossible
to say when it occurs, if indeed it does. When a tree is cut down, the
stump will sprout and with some species produce a tree again. Flowers
blossom in a vase of water; fruits ripen after being picked; seeds may
germinate even after thousands of years.

But we cannot say that the seed is immortal, for it can be destroyed,
a consequence of the fact that it is composite. The animal principle, on
the contrary, being at level II, is not composite; *it cannot be destroyed.*

This gives support to the existence of something which could be referred to as an "animal soul" (interestingly, the word *animus* means "animating spirit") which not only survives the death of the animal, but has existed throughout the history of the species. Such an entity would account for instinct by providing an indefinitely long period for learning and an explanation of its inheritance in the young animals.

But because all animals of the same species have the same instincts, we can say that these animals have the same soul, that is, *a group soul*. In other words, there is presumably a group soul for polar bears, a group soul for wrens, a group soul for the wasp that buries its eggs in larvae, and a group soul for bees. Instinct can accordingly be regarded as a fixed pattern of activity inherited by the species collectively from behavior learned earlier in its history.

It is our business to fill in the blanks which the theory affords, to apply to theory. Here is a case in point. The theory anticipates something analogous to energy at the sixth stage which, like the energy of science, is conserved.

If we at least open our minds to the possibility of a group soul, a number of interesting bits of evidence become available which support the concept.

Especially important is the work of Eugene Marais described in a little book called *The Soul of the White Ant*.* (I was pleased to see that Robert Ardrey dedicated his book, *African Genesis*** to Marais, who was one of the first to observe animals in their natural habitat.) Marais found that workers in a termite colony behave in an impressively coordinated way, for example, moving to repair damage to the colony even when there is no detectable means of communication. And a plate of glass interposed between the queen and the workers does not affect communication, but removal of the queen does. Their termite colony behaves as a single organism, with workers and soldiers responsive to an unexplained communications system dependent on the queen. This is good evidence for what I am calling the group soul.

Again, the coordinated movement of a school of fish or of a flock of birds seems more comprehensible if we assume it to be guided by a

* Marais, Eugene. *The Soul of the White Ant*. Translated by Winifred DeKok. London: Methuen & Co., 1939.
   ** Ardrey, Robert. *African Genesis*. New York: Atheneum, 1961.

coordinating principle than if it depended on *ad hoc* communication between members.

If we now reflect that what Marais called the "soul" of the white ant, which guides the activity of the worker and soldier termites, does so with no apparent physical communication, and that flocks of birds and schools of fish are in all likelihood similarly coordinated, we may wonder whether there is any real difference between these cases of coordination and those of other animals exhibiting instinctual patterns of behavior. Some nonphysical agency seems to be required to explain the termite colony; would not this also serve to explain instincts in general?

Another clue, which I have heard of from several sources but have not been able to establish with certainty, is that in the training of rats the control group appears to profit from the learning experience of the experimental animals, to the despair of the researcher. Even a small "leakage" of this sort would confirm the group soul hypothesis. But since there is no currently accepted theoretical framework into which this sort of thing would fit, most experimenters would doubtless be reluctant to advertise what would generally be interpreted as some error in method, the control group not properly isolated or the like.

One of the most carefully studied and yet mysterious instincts exhibited by animals is the ability of birds to navigate by the stars. The blackcap, a European song bird, annually migrates from Germany southeast to Turkey, then flies directly south to Egypt, returning in the spring. By placing young birds who have not yet flown the trip in a planetarium in which it is possible to duplicate the night sky as it would appear at different parts of the earth's surface at different times of the year, it has been established that the bird navigates according to the position of the stars.

If the sky as it would be seen in Germany is displayed, the blackcap flies southeast; if the sky of Turkey, it flies due south. If the sky as it would appear in France is shown, the bird flies east; if the sky of Persia, it flies west. Since these abilities are not diminished when most of the stars are obscured as if by clouds, the bird must have imprinted on its consciousness not only a very complete star map, but it must have an internal biological clock of great accuracy, since an error in time of one minute would put it fifteen miles off course.

But fantastic as this ability is, it is somehow inherited from the ancestors. The achievement of millions of years of learning and evolution is behind it, whatever the means by which the achievement is transmitted to the young bird.

More recent work on the navigation of birds by the stars provides important additional clues. Emlen* reared young birds where they could not see the stars and found that when the time came to migrate and the birds were exposed to the night sky, they were not able to navigate, demonstrating that the young birds required direct experience, as his next experiment proved, to learn the right stars to navigate by. This next experiment was most interesting—he adjusted the planetarium to turn around the star Betelgeuse instead of Polaris, as though Betelgeuse were the pole star. These birds, when the season came to migrate, behaved as they should were it the case that the earth were tipped so that the axis of the north pole coincided with Betelgeuse.

This brings out the important fact that the birds are able to adjust to the changes in the star map produced over the centuries by the precession of the equinoxes, which gradually shifts the pole through a circle, only returning to a given spot every 25,000 years. Such instinctive behavior, like that of the wasp which seeks out a particular kind of caterpillar in which to lay its eggs, must have been learned over millions of years and requires the type of memory we can anticipate at Level II, that of the group soul.

Can the theory be put to test? I believe it could. I have heard that English tits have learned to open milk bottles. It would be simple enough to isolate young birds so they could not learn from other birds, and see if they could open milk bottles—though I suppose the die-hard geneticist would still claim that there had occurred a milk bottle mutation in the species. Maybe tits just have a fundamental affinity for milk.

* Emlen, T. "Stellar Orientation System of a Migrating Bird." *Scientific American* (July 1975).

XII | Evolution applied to man

We may now come to the question of man. While philosophers have recognized that even basic questions of existence issue from man, and organize their studies around human consciousness, the sciences, in their quest for the laws of nature, have bypassed man and seized on those aspects of reality that are subject to law, to the neglect of those that are not. There is, in fact, no true science of man, nor even of life.

This neglect could be attributed to the complexity of life processes, of which recent discoveries in the chemistry of DNA afford a glimpse, but these discoveries, while of great importance, are mere peepholes into a vista of the mystery of organic nature. The claim of biologists that they have discovered the alphabet of life, when we take into consideration that the DNA molecule that constructs a bacterium contains information that would fill a thousand-page volume, only shows how very much more there is to learn.

But the sheer complexity of the phenomenon is not the only reason that life is not reducible to the principles of mechanics and chemistry. We have developed in previous chapters the basic contrast between a cosmology based on deterministic science and one based on freedom. Since deterministic science is based on the presumption that there is a cause for everything and a predictable outcome, it cannot, by its own tenets, accept free will as a basic ingredient. Purpose and motive must be excluded. Clinging to this principle denies science access to a recognition of life's essential dynamic, by which it thrusts not only against the flow of entropy, but also against any restraint, and creates exuberant variety where necessity would at best maintain a monotonous repetition.

The same criticism could be applied against the much overrated principle of the survival of the fittest which is a valid principle to be sure, if not a tautology, for it is certainly necessary that life survive. But it is an inadequate principle, because the necessity that life survive does not account for the initiation of new forms, for life's continuous exuberance, and its evolution from simpler to more complex.

This brings us to the importance of introducing uncertainty or freedom as the basic constituent of existence. Thanks to quantum theory, this can be done without any basic change in the structure of science, for quantum theory has already introduced the needed reform and, starting with protons and electrons, has pushed the application of quantum principles into the interpretation of molecular bonds. Such bonds were once thought of as static hooks or attachments between molecules. Quantum theory, in contrast, sees the molecular bond as a very active engagement involving electron exchange, resembling two dogs fighting over a bone, rather than a joint held by glue. Important to this exchange phenomenon is its frequency, which establishes a cycle of action. Such a cycle of action makes plausible our speculative guess as to how life begins: that by control of timing the quantum of action is able to build order against the flow of entropy, and so initiates life at the molecular level. The cycle of action, as we have explained, is consciousness.

Specific problems of man

But from the molecule on, we lose sight of the quantum of action, and therefore have no endorsement from physics. Nevertheless, we see life evolving through plants and animals to man. With man, we have testimony of a different sort, our direct subjective sense of freedom of choice. True, this freedom is questioned, but by whom? By the determinist, who speaks with a conviction that is allegedly based on the absolute rule of scientific law. He is apparently unaware that this absolute rule of law was deposed some time ago by quantum theory. Nevertheless, his conviction still stands. Based on what? On rational grounds. Here we refuse to get embroiled; the rationalist is impervious to evidence. Let him go. Perhaps he will eventually discover that he is caught in a circular argument.

The difficult problem is not to prove free will, but to answer the question: how is the monad, a fundamental spiritual principle, modified? In what respect is the "élan" of an animal more evolved than that of a plant or a molecule? The question is difficult, if not impossible, to answer because we are not in the realm of things—i.e., objects that have definite properties. The élan, spirit, or monad is not so much a "thing" as it is a *power*—and it has no measure other than its competence.*

Meanwhile, we have the thesis that the universe is process, confirmed by the evidence for large-scale evolution through stages and grades of organization. Moreover, we have new intellectual equipment which includes the recognition of forces and energies that escape the determinist.

We will in this chapter apply this larger view of evolution to man; and we will show how inadequate is the contemporary view of evolution. We will also indicate the need for distinguishing among several kinds or levels of evolution, corresponding to the several vehicles of which man consists—the cellular organism, the animal body, and the spiritual monad. Only the last is of concern to man, because his cellular and animal vehicles have been inherited from prior kingdoms.

In attempting to treat man in a scientifically respectable manner, we maneuver ourselves into the position of an embarrassed suitor popping the question. Our quest is obvious, but the closer we get to it, the more we take refuge in circumlocution. Scientific parlance becomes circuitous when we try to discuss man himself.

But that is what we want to talk about. To do so, we need to go

---

* If we assume that the equation $E \times T = h$ (Energy times Time equals Planck's constant) holds throughout the arc, and that the time increase in the second half is the same as in the first half, we would have a period of about $1/10$ of a second for human consciousness, close to the alpha rhythm. The time span required to hear a musical note (beats slower than sixteen per second) is close to the beta rhythm. When these are higher-frequency, *more* than sixteen per second, the result is a *sound* (low hum). So too with less than sixteen frames per second in a motion picture, the illusion of motion disappears. However, according to the formula $E \times T = h$, this lengthening of time would be accompanied by a billionfold reduction in energy, which seems quite unreasonable. We are accustomed to references to the still small voice of conscience, but surely it is not that small! I am unable to suggest an answer to this problem—except that perhaps, for the monad, the question of energy is academic once it has learned to control timing.

through a number of anterooms and show our credentials to a number of peripheral "guardians of the throne." Our goal is the evolution of the individual person, a topic that science ignores altogether.

The uniqueness of man: dominion

Let us first consider the objection that science might make against creating a different kingdom for man than for animals. Science has classified man as so similar in anatomical structure to apes that he must have a common ancestor. The "anthropoid ape" is so named (from the Greek *anthropos*, meaning "human being") because of his resemblance to man. Science further supports this view by citing the evidence of fossils for forms of man which were more ape-like than present-day man. There is no questioning the validity of this evidence. Even if man is not descended from apes, he has the body of a vertebrate mammal.

But man is somehow different from animals. We may describe ways in which this difference manifests in his physique: man stands upright, thus freeing his front limbs for countless other uses; his thumb opposes the fingers, making it possible for him to grasp objects; his brain is larger; he is the only animal with buttocks; he is naked; his penis is large (compared with the ape's). Or we may describe a difference in his behavior: he makes and uses tools; he carries weapons; he communicates with his fellows through spoken and written language; he deals in concepts; he reasons and plans; he is self-conscious; he is religious. Such descriptions may be helpful and give evidence that man is equipped to transcend purely animal functioning, but their mere enumeration does not establish the clear categorical difference we require for a kingdom.

What do we require for a separate kingdom? Note that since the kingdoms are cumulative (each one builds on the one before), a common basis is no bar to the distinction between kingdoms. Thus both vegetables and animals are composed of molecules, and animals share cellular organization with vegetables.

But the vegetable acquires a new power in that it is able to build from molecules a complex multicellular organism (it is a manufacturing "plant"). The animal adds the power of mobility. Now it would be beside the point to insist that because the vegetable had a molecular constitution, it was the same as a molecule, or because the animal had a cellular constitution, it was the same as a vegetable. So we can go on to say that man's animal constitution is no bar to a separate status as man.

Nor is this claim made on behalf of human dignity. As a matter of fact, it reads the other way, for as we will show in the next chapter, man in his kingdom is far less evolved than are the vertebrates in the animal kingdom. Man, we repeat, is as far along to his goal of dominion as is the clam to its goal of mobility. (See grid on pp. 86–87.)

Since the dominion kingdom is theoretically established by the symmetry of the arc, the assumption to be made is not the existence of the kingdom, but the *legitimacy of man's placement there*. Be he naked ape or killer ape, we need to discover what it is about man that makes him eligible for membership in a mode of being that transcends the one degree of freedom of vegetable growth and the two degrees of freedom of animal motion. We need to look at the human principle and discover its abstract character.

Emphasis on this abstract character permits us to see the upright stance, the opposed thumb, the greater brain size, the use of tools and of language, and of even deadly weapons, as part of a syndrome, supportive spokes that radiate from the dominion principle. These tools provide not just means for conquering nature but, as is becoming increasingly apparent in modern times, the means for man's own destruction and, hence, a challenge, the challenge to achieve *self*-control, to attain the responsibility of stewardship.

The animal in man

In fact, viewed in this light, man and his animal heritage are seen to be not only separate, but in conflict, a conflict whose goal is the emergence of a working combination, of mobility and direction; the horse of power must be controlled by the rider. The ancients expressed this partnership in the person of Chiron, the wise teacher, portrayed as a centaur, a creature with the body of a horse and the head of a man. The failure to achieve this state was symbolized by the Minotaur, the

monster kept by King Minos in the Labyrinth, which had the head of a
bull and the body of a man.

But let us return to observable facts. In the astounding variety of
animal shapes as compared with the invariance of man's shape, there is
the hint of a difference that is fundamental. The vertebrates alone,
from eels to elephants, display enormous variation, whereas man is
inevitably the same two-legged, two-armed creature. Of the thousands
of species of mammals, the tens of thousands of vertebrates and the
hundreds of thousands of species of all animals,* the entire range of
mankind occupies but one species, *Homo sapiens.* In the animal
kingdom it seems as though nature is exploring the world of shapes; it is
creating a tool kit. It invents a battering ram (a rhino), a prehensile,
flexible tube for lifting (elephant), a nibbling machine for cutting down
trees (beaver), a drill for getting insects out of trees (woodpecker), a
high-frequency sonar device (bat), a detector of infrared radiation
(owl), and hundreds of other ingenious forms for coping with the
problems of survival.

But when it comes to man, nature changes the emphasis. It makes
man a generalized user of tools. This is a shift of an abstract and definite
nature, a shift that is just such as is required for a kingdom. Such a shift
could not be accomplished (or let us say we cannot conceive of its being
accomplished) in any other manner than by settling on some one animal
form, and letting this form create the tools that formerly endowed the
separate species.

By making a tool a separate thing, man achieves a tremendous
advantage. He may wield a club, throw a stone or a spear, put on the
fur coat of an animal, even fashion a drinking cup from a horn. In
doing one of these special functions, he does not give up the others, nor
does he incur the limitations that animals have to put up with in evolving
hoofs or claws or tusks. The fantastic variety of bills that birds have
developed, which confine them to quite specialized feeding habits,
shows that evolution with animals can go only so far.

So man surrenders the more immediate advantage of a special shape
or of making his limbs into special tools, and instead uses his otherwise

* Tracy I. Storer in *General Zoology* (New York: McGraw-Hill, 1951) accounts
for 4,400 mammals, 40,600 vertebrates, and a minimum of 800,000 species of
all animals.

ineffectual hands to fashion tools, weapons, habitations, and vehicles to meet the requirements for survival, but he does even more. He begins to disconnect from intimate partnership with and participation in nature.

This is a dangerous experiment. We are familiar not only with the warnings of the psychoanalyst that man must not cut himself off from his lower nature, but we are now hearing the voice of nature itself, as pollution, poisons, and other side effects of progress make themselves felt.

This warning, of course, must be heeded, but the solution is not in specialization, as it is for animals in the horse's legs and teeth, in the elephant's trunk, the tiger's claw, etc., but in disconnection, not just from machines, but from the habits and "mechanism" of the psyche and even the "mechanism" of society. We should rule mechanisms, not be ruled by them.

This need to throw off compulsive embodiment, to disconnect from specific function, to become universal, a total being, is the role of man. In the ability to use different tools, to function in different ways, he contrasts with the animal, which is committed to one element, to one diet, to one habit pattern, and therefore is that function.

Thus man can *have* what animals are compelled to *be*, and man's being is freed to take on greater challenges.

This principle carries the key to the steps of evolution in that each power employs (or has) the power that precedes it, *viz.*:

Particles have intrinsic freedom (action/light).
Atoms organize nuclear particles (to build atoms).
Molecules combine atoms.
Plants organize molecules (to build cells).
Animals feed on vegetables (as food supply).
Man uses animals.

Man's "use" of animals or of the animal power is not only in the breeding and domestication of actual animals, but in the use of machines to provide mobility and other functions. When he uses horsepower, it doesn't have to be with a horse. In this way man gains the advantage of the specializations to which the animal is dedicated without incurring the penalties. Thus man may ride a horse without giving up his hands; by making an airplane he can fly without the severe anatomical restrictions that the bird must meet.

We thus describe man by contrasting him with animals. We observe that the "being" of an animal is properly expressed through some characteristic mode of behavior: it is fox-like, elephantine, leonine, bearish, ant-like, waspish, worm-like, whereas the being of a man is freed from this specific character. If a man behaves like a mouse, eats like a pig, works like a beaver, he is doing so by option, not by necessity, and because it is an option, he is criticized or praised for his behavior.

This leads, then, to a final and rather difficult fact about man, which is that, while we know perfectly well what people are, there is no common distinguishing characteristic of all people. They are, of course, bipeds, vertebrates, warm-blooded, etc., but this is a heritage which, as we have already argued, is carried over from the animal kingdom, much as animals carry over cellular organization from the vegetable kingdom.

This means that it is impossible to ascribe a specific nature to man. The goal of evolution is that which transcends limitations, and, since to define is to limit, we cannot ascribe definite attributes to man.

The goal of dominion

The dominion kingdom has as its ultimate goal the evolution of unlimited being, ultimately of God, a traditionally ineffable existence, inexpressible, unspeakable, but the notion of God is also synonymous with the supreme absolute, something far beyond man and not pertinent to the intermediate stages of dominion. What is appropriate is the ineffability, for whatever the limitation of characteristics which attaches to a person under specific circumstances, or in a specific action, we may in theory expect that it can be overcome. We expect the person to be capable of being other than or more than he was in the action in question.

I emphasize this point because it is essential to our theory, which anticipates this "open" or unlimited area. The second level* is by nature

* Particles (or energy) are infinite because they cannot be destroyed; atoms can (atomic bomb).

infinite, as the third is finite. The open and unlimited quality of the first level is not simply infinity in the sense of "without end" (in mass, temperature, energy, etc.); it is doubly infinite. It is infinite "no-thing-ness."

Rather than struggle to define the undefinable, let us take the opportunity which we now have to know this idea directly. Aside from your body, what are you? What am I? If we find it hard to allow the existence of something that has no qualifications, how can we allow our own existence? If we are a monad that was once in a molecule, in a cell, in an animal—what is this monad? What was your face before your parents were born?

With this preparation, we may now hear the evidence from evolution as it is described in the texts. The following presentation will already be familiar to many readers, but it is my hope to show that it has been put to a number of uses which do it no credit. It has been used as propaganda,* for cover-up, and for rationalization in ways that suggest the errors of which teleology and purpose are customarily accused. Furthermore, and most important in the present context, it has been improperly applied to man.

A résumé of showcase evolution*

From fossil evidence we know considerable detail about the evolution of a number of animals. Let us take the horse, for example. The evidence for the evolution of the horse from a five-toed ancestor (the eohippus, a creature about the size of a dog) exists in the form of fossils of varying age. It is possible, by placing the fossils in a sequence beginning with the oldest, to find the leg of the five-toed eohippus gradually lengthening and the number of toes reducing.

Eohippus	5 toes	
Protorohippus	4 toes	
Misohippus	3 toes	(side toes touching ground)
Protohippus	2 toes	(side toes not touching ground)
Equus	1 toe	(splints of 2nd and 4th toes)

* To contest literal interpretations of Biblical accounts.

Similarly, the elephant has gradually evolved a trunk. Such instances are the "showcase animals" of evolution; it is assumed that other creatures and other functions evolved in similar fashion. But let us note *how* this evolution takes place. Primarily, it is dependent upon *specialization* and the usefulness of the specialized members. The horse specializes in running and, as the land rises and food gets scarce, he has to roam farther to get enough food. Presumably, he is set upon by predators and has to be able to escape. Hence, through survival, the modern horse has evolved legs suitable for rapid travel over land, teeth for feeding on grass, etc. At the same time, the capacity to pick up grubs and climb trees, which the less specialized opossum retains, has been sacrificed.

This example of the horse's leg is considered the prototype of all evolution, but it has a number of limitations which read against it as the solution to other evolutionary problems.

To describe the problem of evolution more fully, we must realize that selection is not the only factor. For selection to operate, there must be variation, and the source of this variation has always been a problem. The older naïve interpretation assumed that things just naturally varied. The ears of corn varied in length, and if you selected the longest ears for seed, you would get long-eared corn. This is only partly true, for while you can get in this way a greater proportion of long-eared corn, you will not get longer ears than were available in the first place. You have purified the breed, but you've not created anything new. In addition, that mechanism which can ensure the reproduction of millions of generations by its very nature must deny random variation; the rules that would ensure a sound product exclude deviations. The printing press can be expected to duplicate a book, but not to write new books.

In the face of these objections, the Darwinian theory (of evolutionary survival) would have collapsed were it not for the rediscovery of Mendel's law and the modern modified version of De Vries' theory of mutations. De Vries held that the chromosomes themselves could be altered irreversibly by cosmic radiation. Such changes would be transmitted, and while most would be undesirable, some would be desirable. Selection would eliminate the undesirable changes. The survival train was back on the track and running again.

Now I have no wish to discredit this theory of mutation due to cosmic rays. It is probably the correct explanation of many phenomena, and there is no doubt such mutation does occur (not to mention that creation of a new species by mutation due to cosmic rays has a familiar ring, suggesting, one is tempted to say, the Virgin Birth). But to account for all problems of evolution in this way is absurd, and does no credit to science, which on this touchy subject has been more political than scientific.

To see the limitations of using the horse as the prototype for all evolution, let us fill in the picture. We must first grant a gradually changing climatic condition which puts a premium on mobility; there must be a need for the grazing animals to range farther and farther to have enough food. Under such circumstances long legs are an advantage, and each increment in length of leg has survival value. The dog-like five-toed eohippus, over a period of fifty million years, continually pressured by necessity to move about rapidly, evolves into the modern horse.

Now apply this to the evolution of the bird. The bird has a greater task than the horse; it must make a jump to a new kind of existence, live as it were in a different element; it must invent. The eohippus, to become the modern horse, did not have to innovate; he was already running about in search of food, and his evolution into the modern horse was a consequence of doing better what he was already doing. The creature that was to become a bird had to give up the use of two perfectly good feet and dedicate them to evolution into wings. This idiosyncrasy would have had no survival value until the creature could actually fly, and this would have been only hundreds of mutations and millions of years later, for flight depends on a number of simultaneous design changes: hollow bones, increased blood temperature, feathers, and many more. Mutation could conceivably have produced these changes separately, and over a long period of time, but it could not have produced them all at once, unless you believe that monkeys could write Shakespeare. And until all changes had been made, the bird *could not fly* and hence could not survive better than four-legged competitors. So the evolution of the bird is not just a more difficult kind of evolution; it involves problems which cannot be solved by the mechanism that worked for the horse—increased specialization.

An equally puzzling problem, which a superficial consideration might dismiss as similar to the horse's legs, is the evolution of the human brain.

What makes the brain different from other evolutionary devices for coping with survival is that the brain develops by *not* specializing. Its value depends on being able to apply the experience gained from one problem to *another problem*. If the brain evolved by specialization, then children would be born speaking English or knowing solid geometry, just as the horse is born able to walk and run almost immediately. Indeed, the evolution of brain or mind is such a different problem from that of the horse's leg or the elephant's proboscis that it may not fit at all into available niches, and may require the type of evolution which we will discuss at the end of this chapter.

Finally, the most comprehensive shortcoming (if we can use such a term) of the current theory of evolution is that it does not account for jumps to a higher level of organization.* As has been pointed out, there are many examples of creatures, low on the evolutionary scale, continuing to survive after hundreds of millions of years. The horseshoe crab is the most primitive of arthropods, and yet is doing very well. The shark is classed below the true fishes, yet it holds its own despite the evolution of higher forms; and there are thousands of primitive life forms which survive well today. Why then is a leap to higher forms necessary? If it were a case of evolve to a higher level or perish, the lower forms would have dropped out. In addition, the leap to a higher level of organization cannot be explained genetically any more than could the accidental mix-up of parts turn a carriage into an automobile, or a telephone into a radio.

The problem presented by those leaps from one level of organization to a higher one is not even considered by current theories of evolution. Its treatment, in fact, takes us right into the theory of process we are presenting. It cannot be studied by examining the details of zoology; it requires a more comprehensive view, one that recognizes the problem of leaps to higher types of organization in other areas, in the development of molecules, or of the shells of atoms (for atoms and molecules, as is evident in the grid, evolve through the same substages of organization as do animals, yet survival of the fittest can hardly apply to atoms and molecules).

To sum up then, there are important respects in which the current theory of evolution (survival of the fittest) is inadequate, and we should

---

* LeCompte de Noüys makes this point in *Human Destiny*, New York: Longmans Green & Co., 1947; David McKay, 1947.

invite some new approach. This is offered by our theory of process. We have already (in Chapter XI) indicated how the hypothesis of a group soul would account for instinct, but this does not exhaust the possibilities.

Genetic and instinctive evolution

If we return to our four levels and note their relevance to the problem, we can account for at least two sorts of evolution: genetic evolution and the development of instinct.

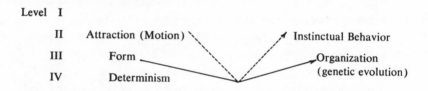

Level I

II Attraction (Motion)   Instinctual Behavior

III Form   Organization

IV Determinism   (genetic evolution)

The problem of any evolution is to achieve control and reach the volitional side of the arc. For genetic evolution this is done by the test of survival, for development of instinct by trial and error. While the former is recorded in the DNA and the latter in the group soul, the *proving ground* for both kinds is level IV.

The distinction between genetic evolution and instinct could be likened to that between the design of an automobile and the driving habits of the operator. The design evolves as different models succeed one another, self-starters, balloon tires, four-wheel brakes. The driver accumulates experience with the years: he learns turn signals, traffic rules, to anticipate the other drivers' intent, where his friends live, and so on. Both kinds of evolution occur and are necessary.

Group soul

In the last chapter we introduced the concept of a group soul as a hypothesis which would account for a number of inexplicable phenomena, including the elaborate instincts of animals. In this chapter we are emphasizing evolution and will consider the group soul concept along with other theories. Our goal is the evolution of man, but to reach it we have a certain amount of spade work to do.

For example, legitimacy. The group soul concept might be ruled out on the grounds it is inconsistent with scientific principles (whatever these may be), but it is not inconsistent with the theory of process; in fact, it is implied by the theory we are presenting. Even further, I believe it is implied by the basic principles of science.

Because the group soul comes into existence with the animal principle, it is level-II and hence of the nature of an *energy* which, by the principle of the conservation of energy, *cannot be destroyed* (as can form, meaning an assembly of parts such as atoms, molecules, or cells).

The group soul concept follows from the theory of process because the theory of process requires that the animal principle be immortal. (Just as science *requires* that energy be indestructible.)

The group soul concept does not replace genetic mutation, but it can implement it. Applied to the evolution of the horse, the mutations which give rise to long legs could well be assisted by, and might even require, behavior which takes advantage of the long legs (running). In other cases, such as the termite colony observed by Marais, the group soul is the only explanation. In fact, based on the genes alone, we could not distinguish the queen from the worker termites, or the soldiers from the drones; this differentiation is based on group need, which may be equivalent to what we call group soul. When the queen is killed, a worker termite is given the special food that transforms it to a queen. Since the queen and the other termites—workers, soldiers, and drones—have the same DNA, there are factors other than DNA operating. These depend on the needs of the whole and thus suggest an organizer body which is a kind of group soul independent of its individual members. It is evident that there are new worlds of connection, new additions to science in the possibilities which unfold.

The group soul concept, it might be mentioned, provides a mechanism for the "inheritance of acquired characteristics," alternative to that given by Lamarck, who first proposed such inheritance as an explanation of evolution. His theory is very much out of favor today, primarily because the accepted mechanism of cell reproduction (DNA) cannot account for either the learning or the inheritance of acquired characteristics (acquired characteristics would include behavior learned by trial and error). That is, because the cellular material responsible for sperm and eggs is isolated from the rest of the organism early in life, the behavior of the parent cannot be imprinted on the germ cells and hence cannot be inherited.

This conclusion may prove somewhat hasty, for while it is true that the chromosomes cannot be affected by the behavior of the parent, it is by no means self-evident that the cytoplasm (the cell content that is not in the nucleus, the DNA) could not be affected by the behavior of the parent. In fact, this would be the explanation of recent work which has shown that planaria worms inherit the education of the parents. The inheritance of memory in planaria can be explained as carried by chemical substances. I therefore cannot call on planaria to support the group soul theory, but I can use the example of planaria to illustrate that there are two possible kinds of inheritance: the accepted one attributed to DNA, and another that can account for planaria. The former depends on form (level III) and the latter on substance (level II).

These two principles, form and substance, are so basic to our theory, and so universal, that we can almost make it a rule that where one is, the other is. In the case of planaria, substance is involved in the literal sense that the blood stream carries substance that was once part of the parent. Even more so in the case of young worms fed on worms that had been conditioned to avoid the light. The new worms learn more quickly (to avoid the light). One theory is that the conditioning experience creates molecules which remain in the blood.

Our theory goes further in that it postulates a more ultimate substance, *psychic energy*, which in animals is the group soul. What the theory cannot say is at what stage in animal evolution the group soul emerges; the very primitive animals, sponges and coelenterates, can hardly be said to have instincts; their mobility is too limited. On the other hand, with insects, colonial insects especially, instinct is elaborate. Ants and bees behave in a way that suggests a highly developed group soul. Thus we must suppose that group soul, like the animal, is evolving; indeed, it would be more correct to say the animal group soul is the animal, and is evolving in conjunction with its vehicle, the cellular organism.

Can we find other support for the group soul concept? I myself was impressed by a small book on the subject written by the theosophist Annie Besant. But the group soul concept has always been part of the tradition of "primitive" people, who, as a matter of fact, are in a far better position than modern man to have a correct understanding of wild animals since their existence depends on it. Having considered various theories which apply to the evolution of animals, we can now ask how they might apply to the evolution of man.

What evolution? How do we know there has been any evolution of man? Granted, of course, that civilizations rise and fall, granted that science and technology have made an increasingly rapid progress in the last two hundred years, granted that more people than ever before are receiving the benefits of education, vaccination, and medication, granted anything you like, equality for women, biodegradable containers, colonies on the moon—*has man evolved*?

Man's peculiar evolution

For all the discussion of evolution, including the question of man's ancestry, there has been no intelligent attempt to apply the concept of evolution to man himself. Most of the ammunition went to proving man was descended from ape-like progenitors, but so what? Granted that man has an animal body (and we have explained why this does not mean he belongs to the same kingdom, any more than the cellular constitution of animals means that they are vegetables, or the molecular constitution of plants means they are molecules), why should he not be ape-like? In any case, the real issue never seemed to come up.

That is, what effect, what *meaning* has evolution had for man? There is certainly no detectable difference in human physique in recorded history. Ancient sculptures show that. And though present humanity may have evolved from such characters as are found in the files of anthropology, the Peking man, the Java ape man, the Heidelberg man, and the Neanderthal man, the steps in this evolution are by no means proved, especially since contemporary with some of these was the Cro-Magnon man, who was quite a fine specimen, taller, and rather more handsome, than modern man. Furthermore, had man's evolution been directed toward body changes, it would have produced as many different kinds of body as animal evolution has with the 800,000 species of animals. How can Darwin's theory on the origin of species by means of natural selection have any pertinence to human evolution, when all mankind is but one species?

How then has man evolved and how will he evolve in the future? Body changes not being significant, we might suspect he would evolve a better brain, but has he done so in the past? Are our present philosophers, our writers, our sculptors, better than the Greeks? Our technology may be better, but this is not necessarily because the engineers are more intelligent; it may be because of the cumulative effect of the past.

Weakness of Darwinian evolution for man

We should not overlook the fact that for man the survival mechanism on which Darwinian evolution is dependent breaks down, and goes into reverse! In modern civilization, for example, the most fit tend to have fewer offspring. In fact, in India the more highly evolved individuals in many cases dedicate themselves to a celibate life, and there are other ways in which civilization, by preserving the sick or the mentally unfit, tends to annul the selective factor in man as a species. This again stresses that there is another type of evolution for man as an individual.

The crowning touch is that according to the genetic theory, our struggle with adversity—our wars, our trials and tribulations, our education, our search for truth and for the good—because it does not affect the germ plasm, has no effect on the genetic evolution. It is entirely wasted. In terms of present genetic theory, we might just as well omit civilization, life, work, careers, just put our seed into bottles and send it to the breeder. We ourselves are not even as useful as cattle, who do produce meat and dairy products.

These are not idle speculations; they are the considerations into which we are forced by the stampede to turn the present incomplete notions of science into dogma.

Inadequacy of group soul concept for man

Nor can we do any better by calling on the group soul concept, so helpful for animals, to help in accounting for man. If anything, with man the group soul must be operated against. Man evolves not only insofar as he is able to "break away from the herd," which is our unconscious recognition of the group soul, but also insofar as he "overcomes his animal instincts," that is, goes against the behavior pattern that for the animal would be an optimal if not necessary solution to its survival (see pp. 128–129).

To get to the problem of man, we must look at humanity not as a species, but as made up of individuals. When we do this, the technical unity of *Homo sapiens* gives way to great variety. Have you ever known two people exactly alike? I have known several pairs of identical twins whose appearance was so similar it was almost impossible to tell them apart, yet in character they were altogether different, as different in fact as are people who are not identical twins. There are many respects in which people can be alike, sex, color, height, education, cultural background, etc., and twins might be identical in these respects and quite different in their character. It is the latter that we should hold in mind when we are speaking of human evolution. How does a person's total character evolve? The color of eyes, hair, etc., is genetic, inherited in the germ plasm, as are all anatomical details, and we might speculate that the climbing instincts of small boys is an inheritance of the kind of instinct and behavior which becomes important to animals. But how do we account for the unique character traits that make, say, a Mozart?

The evolution peculiar to man: the individual soul

Here is where our theory offers a third type of evolution. In plants we have the evolution of the cellular organism traceable to the DNA. In animals there is added the behavioral evolution which is made possible by what we call the group soul. In humans we have a yet different kind of evolution, that of the *individual* soul.

An individual soul is not a group soul, but like the group soul it is immortal. Like the group soul, its experience is cumulative (the "soul" in both cases is the memory bank drawn from former existence in which lessons were learned), but the individual soul has a different task from that of the group soul. This is to transcend, rather than to repeat or react to, its previous experiences. In other words, the animal is exploring and exploiting to the fullest the way of life that its instincts open up—the beaver builds dams, the anteater confines its diet to ant hills; neither would think of going to a psychoanalyst to get over his mental fixations, or to broaden his education by the study of Sanskrit. Yet man is under an obligation to take on greater challenges; we can even say that he wants new kinds of experience, that he may enjoy a routine for a while, but that it is normal for him to make basic changes during his relatively long (by animal standards) lifetime. The pleasures of being a football player lose their appeal in the mid-twenties at just about what would be old age for a horse.

In another sense, man may stick to one skill for more than one lifetime. For example, Bobby Fischer, like other chess champions, exhibited early in life such an exceptional ability in and *inclination* for chess that he must have learned chess in previous lives. And note the importance of incentive. Bobby Fischer is highly motivated. Even if we accept what I am stating is impossible, and that chess playing *behavior* could be transmitted by DNA, DNA could not transmit his powerful incentive—which keeps him continually thinking and dreaming chess, even to wanting to smash his opponent's ego, an issue which would not concern a computer. General Patton is also interesting in this respect not only because of his conviction of previous lives as a general, but because of his courage in publicly stating this conviction.

Against modern opinion, which would rule out not only preexistence of the soul, but the soul itself, one could cite Plato, whose doctrine of the soul's immortal nature was taken over and later questioned by Aristotle. But it is hardly correct to credit Plato with this idea which was general in Egypt three thousand years before Plato drew breath, and is in fact current in most religions, notably Buddhism, Hinduism, and in almost all primitive religion. The soul's preexistence was dropped from Christianity only in A.D. 553 by an edict known as the Anathema against Origen, pushed through by the Emperor Justinian while the Pope was in jail.*

I confess myself puzzled as to why it suits the Christian church to deny the soul's preexistence; such denial certainly makes the soul's immortality (which the church affirms) less credible, for how can that

---

* Head, Joseph, and Cranston, S. L. *Reincarnation*. New York: Julian Press, 1961; Theosophical Publishing House (Quest Books), 1968.

which doesn't die be born? I rather think that the importance of individuation to modern civilization is the factor that foists upon us this phantasm of a piece of twine that has only one end, the Christian soul. (An endless twine is no more incomprehensible than is time itself.) "For that which is born death is certain," as the Bhaghavad Gita puts it.

Yet we must be careful not to overemphasize the soul. Not only is it subordinate to the spirit (the seventh and first principle), but in a sense it does not yet exist for most persons. We have several times mentioned that man is not very far along in his evolution through the seventh stage, and hence for man soul is second-substage rather than sixth; it is involuntary and unconscious rather than voluntary and conscious as it would become in substage six. This not only accounts for its obliteration in modern life, but is consistent with the most important teaching of Christian religion (and of other religions), the Virgin Birth—but that would take us beyond the scope of this chapter.

What we should complete before closing this chapter is to sketch in what we described as a third type of evolution, one having to do with man.

Level   I   o   o   ooooooooo   Man's evolution

   II   --------   Animal evolution

   III   ————————   Plant evolution

   IV

Just as plant evolution requires DNA (level III) and animal requires the group soul (level II), man requires something intrinsic to level I which is almost better described as the divine *light* than by words such as consciousness or understanding, for it is the light of recognition that leads to the "turn," just as it is the capture of a photon that procures the negative entropy of plants.*

---

* The capture of a photon by chlorophyll exemplifies the turn, which makes possible the evolution of higher kingdoms on the right of the arc. This should not be confused with the evolution of DNA, which is *one* step in this evolution, but requires the interaction of *three* powers: form (to create the drawings), survival (to select the fit), and organization (to manufacture).

The three types of evolution and their dimensionality

In view of the foregoing discussion, then, we have three types of evolution. First there is the evolution of the genotype, which is effected through DNA. This is the evolution recognized by current science, and produces the cellular organism.

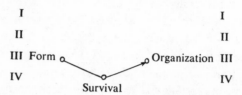

I   I

II   II

III Form   Organization III

IV   IV

Survival

It is an evolution of the *design* of the mechanism; the blueprint which it starts with as a form at level III (stage three), tried and tested at level IV, and if not eliminated by failure to survive, produces the prototype at level III (stage five).

The process could be likened to the evolution of a mechanical device, an automobile or airplane; the form (level III) is tested at level IV and the feedback evaluated. Where weaknesses are indicated, corrections to the form are tried over and over until it is adequate for production. When this occurs, the process moves on to the organization or growth stage, and a new species comes into existence. This is just how plants evolve, via the cellular organism.

The second type of evolution begins only when there is the possibility of choice, and hence *behavior*, for without choice there can be no behavior. The behaviorist might deny this, but how does the animal learn to go through the maze? The right choice must be reinforced. A stone, which has no choice, could not be taught to go through a maze. This second type of evolution produces animal *instinct*. It begins at level II, where attraction causes movement toward or away from the various things in the creature's environment: food, light, warmth, water, other creatures, a mate, etc. Just like the first type (of evolution), instinct is tested at level IV, but its "survival" in this case is better described as trial and error: the creature makes exploratory motions and receives reward or punishment which reinforces the appropriate behavior.

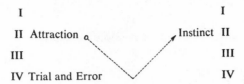

I

II Attraction

III

IV Trial and Error

I

Instinct II

III

IV

Where the correct choices to achieve the goal have been found, a behavior is learned, and this becomes what we call instinct. There is nothing difficult in this explanation except to account for a way of passing on this instinctual behavior to the progeny. This is impossible by the blueprint mechanism. As we pointed out earlier (Chapter XI), it requires an immortality or conservation of the desire energy of the animal, group soul. The possibility of a group soul is implicit in the notion of the second level, which is such that it does not end in time. Like mass and energy, the "substance" of the second level—which is the "psychic" or desire energy that animates the animal, which in fact *is* the animal—is evolving in time and does not die with the individual.

The difficulty we have in accepting this notion of immortality arises from the fact that we identify the animal with the cellular organism which does die and could not even theoretically be immortal because it is composite. Of course, the DNA is destroyed when the individual dies, but it survives through the other members of the species. That is to say, the DNA is effective through any member of the species, and for that DNA to die entirely, the whole species would have to be wiped out. Even in such a case, the DNA might survive in seed form; for example, seeds buried in Egyptian tombs thousands of years ago have been found to grow and reproduce.

This raises an interesting idea. Mammoths, a long-extinct species related to the elephant, have been found buried in the frozen tundra of Siberia in such excellent preservation that their flesh is still edible. Perhaps their seed could be fertilized and transplanted to the womb of an elephant to produce a living mammoth. Even if this experiment were successful, we might then find that there was no mammoth group soul to animate the cellular organism.

We are faced with the quite serious challenge of recognizing the animal for what he really is—that an animal is not only a particular lion or zebra, but a representative of the lion or zebra *species*, and the species is the group soul of that animal which draws on the experience of the species for its inherited instinct and, by its own experiences, adds to the species' experience.

The third type of evolution is possible for man.

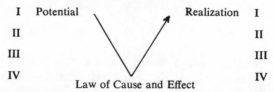

I Potential   Realization I

II   II

III   III

IV   Law of Cause and Effect   IV

This evolution contrasts with that of the animal in that the possibility of *understanding* the law is added to the trial-and-error syndrome of the animal; this understanding offers the possibility of a much more rapid evolutionary advance. This third evolution is still subject to the need for individual effort, for pain and travail. But it also affords the joys of creative endeavor and the enjoyment of aesthetic sensibilities.

The solution is no longer cloaked in the darkness of inherited instinct which depends (like the mating instinct) on blind obedience to sensory clues that vary with the season; it invites the self to explore the world both of the senses and of the abstract reason, the heights of art, the emotions of love, the discovery of truth.

Like the plant and animal evolution, this third type has its laboratory at the fourth level, where the law of cause and effect affords the opportunity for recognition of truth, for the enlightenment that is the goal of this evolution. Once recognized, the law of cause and effect makes it possible to bring an effect about, to make determinism serve will. Thus the universe is a school in which the monad learns.

Additional information

The dimensionality of the three types of evolution:

Taking advantage of the principles which the theory of process provides, we can supplement the argument just given by reference to the dimensionality which distinguishes the level.

III   Form   Organization

The first, or genetic, type of evolution involves DNA and the cellular organism; initiated in the third level, it is corrected at the fourth. The third level, that of form, requires two dimensions (much as the blueprint requires a sheet of paper).

DNA, it might be argued, is linear. But this is not really the case; though the *storage* of DNA is linear, much as the magnetic tape which stores a concert is linear, the information is available as it is in a book, to be used at the appropriate time. What we currently know about DNA is only that the information is there; we have not yet learned how the appropriate information is applied at the right time. Thus there is a hidden dimension in DNA, one not yet discovered. If this is not convincing, we have the more basic requirement that any information is two-dimensional, as it defines concepts and operations.

II   Attraction   Instinct

The second type of evolution, the instinctive, occurs at level II. It is an evolution of behavior. Therefore we require *time*, which is one-dimensional. We have already pointed out that level II deals with values (success is good, failure bad) and values require one dimension only (because more than one dimension would not determine a value unambiguously). We might add that because it is confined to time, the animal is able to form the associations that consolidate its correct choice with an instinct pattern. That is, its successful choice and the reward come at the same time, and therefore the animal associates the reward with the direction of choice.

I   Light   Realization

The third type of evolution occurs neither in two-dimensional space nor in one-dimensional time, nor does it occur in the three dimensions of the physical world, for this (the physical world) is the testing ground of all three evolutions and contains no activity of its own. It is a screen rather than a source; it eliminates the unfit, or verifies the correct behavior.

The third evolution arises at the first level and ends when its goal is achieved. Like the others, it requires the fourth level for its laboratory, but its experimentation does not occur in either space or time. It is instantaneous; it occurs in the *present*. The present is not a part of time, because it is always the present. Time is the shadow of the present and derives from it.

Now we still need to bring out the peculiar potency of this evolution. Primarily it is *recognition*, recognition of a principle, realization of a truth, reconciliation of a duality, *satori*. It is at once the privilege of man, and the formative principle that enables man to evolve. It is an evolution that is realized through man's potential divinity.

XIII | The substages of dominion

We now propose to describe the stages through which man as an individual monad evolves. Such evolution does not concern his physique, which is an inheritance from the animal, nor is it concerned only with intelligence. It is, rather, a combination of all his human faculties: intelligence, emotion, intuition, will, and, above all, the integration of all his talents into a single whole.

Moreover, man's environment, as becomes evident with increase in population, is of a different nature from that of the animal's. Animal evolution consists primarily in developing a physical organism that through its mobility can obtain food and by its internal organization can transform food into available energy. Man begins his evolution with the animal vehicle already perfected. Food supply and other problems must be solved, but even with primitive people man's focus of attention directs itself to another kind of effort than that of direct interaction with nature. He becomes a member of a tribal society and occupies his time with ritual in a variety of forms.

This brings out the importance of civilization, or of the social state, in man's evolution. In what follows we must necessarily acknowledge man as a social animal. However, as we will endeavor to show, it is the self, and not the state, that is evolving. The state or the civilization in which man finds himself is as necessary to his evolution as is nature to the animal's evolution. Nevertheless, what is evolving is not the state, but man as an individual being.

This might seem to imply that we are backing the individual versus the state, but that too would be a misrepresentation, for, as we will further endeavor to show, individuation is itself but one part of a more comprehensive process in which a transcendent goal is attained. The goal is to reach a higher development than present man, whether we call this superman or to become "wise as gods." The process by which that goal is reached requires the interaction of persons with one another, and of persons with the state, but neither the centrality of personality nor the communality of the state is its final resting point.

Before proceeding with our subject, the substages of dominion, we should caution against too literal an interpretation of the division of process into stages. Stages (or substages) are not different things, but different phases in the development of the same thing. Perhaps we should say different vehicles for the development of the same monad. As with beads, they may conceal the thread that holds them together.

So when we refer to different periods of history (early man, the Greeks, modern man) as stages of dominion, we mean that in these periods a preponderance of persons were at a certain stage—not that "civilization" was at this stage, because, as we said, civilization is the vehicle, not the end product.

First substage

The obvious instance of first-substage man is the primitive hunter, before agriculture introduced the tribal community with its division of labor, its priesthood, its dependence on the seasons and need to determine the correct time for crop planting. But I'm not too sure that this is the answer; most early peoples describe themselves as having been taught by the gods, who showed them how to plant crops and weave, and brought them music, writing, and the other arts. There is also the theosophical approach, according to which the earliest man was not yet in a physical body, as if to emphasize that man had a quite different origin from that of the animals. Finally, there is an account which describes the "lords of Venus" first preparing the higher apes for take-over by human souls. As the story goes, some of the souls refused to inhabit the ape bodies—a tragedy because of its effect on the apes. I'm rather taken by this account, particularly when I look at the expressions on the faces of apes.

Second substage

The general name for the power at this stage is binding. In terms
of dominion, and in terms of human evolution as distinct from
animal evolution, this would be the stage before the emergence of man
as a self-conscious and self-determined individual. It would be man
functioning as a member of a tribe or civilization in which his actions
were completely prescribed by tribal customs, or by king or leader.
To some extent, all groups partake of this behavior, but we must
remember that the later stages, by virtue of the cumulative principle,
always retain the powers that have evolved before. So that where
modern man may still retain group behavior, he can also be
self-determined. He may change his allegiance, and this allegiance, even
if not completely voluntary, is not the unconscious and implicit
allegiance of the collective society of substage two. Seen in this light, the
political philosophy of communism is a reversion to this earlier stage of
development, and we would read the changes that have occurred in
communism, which carry it ever further from its original aims, as due
to the fact that the present stage of human development has passed
beyond that of simple collectivism.

Again, the "organization man" as a contemporary phenomenon is
not so much an instance of collectivism as it is a basis or backdrop, a
prior condition, for self-determination. The individual is able to find
himself only by questioning the ethos or mores of the group, by
disobedience, as Genesis puts it. Again, the organization man of today
is a far more complex and self-conscious entity than is apparent at first
blush. The term refers not to a child-like zombie who lives entirely
according to a sacred ritual, as we may suppose primitive man to have
done, but to a member of an intensely competitive society who employs
a uniform codified behavior to better himself or to obtain status, only to
shift into another garb when the occasion presents itself.

All of these conflicting considerations vanish when we go back in
time and consider a primitive civilization in which collectivism was
unalloyed with elements of individualism. Take, for example, the Inca
civilization which Pizarro encountered in the conquest of Peru.
Contemporary accounts describe a civilization of some twelve million
peons who cultivated their terraced gardens, chanting as they worked,

devoted to their king with a complete and unquestioned faith. Everything was regulated and set forth in prescribed ritual; there was no need and no occasion for initiative. Yet this very perfection rendered the civilization extremely vulnerable to attack, for how else can we explain how Pizarro, with his band of some one hundred eighty soldiers, conquered a race of twelve million? For when Pizarro captured their king, the collective entity collapsed.

We could liken such a civilization to the termite colonies studied by Eugene Marais, to which we referred in our discussion of the group soul in Chapter XI. He found that when the queen was removed, the whole coordinated activity of the colony broke down: the soldier termites ran about confusedly, the workers stopped repairing, the colony disintegrated.

The reasons for the breakdown cannot in either case be fully explained. The point that is important is that the level of development of the individual units—the peons of the Inca collective society, or the termites in the colony—was not sufficiently advanced for them to take command and resist intrusion when the leader was removed. In a more advanced civilization there would be enough self-determination at the level of the common man to organize resistance. We have many illustrations of this phenomenon in modern history: the American and French revolutions, the underground resistance in the countries overrun by Hitler, the organized resistance encountered by Mao in China in the 1960's, the warfare in Southeast Asia.

Other interesting implications can be drawn from the evidence. In the case of the Incas, there appeared to be no sense of the importance of individual survival. Once their leader was seized, resistance collapsed. This suggests that morale is involved, and such morale has its source for the collective man not in himself, but in someone or something beyond himself and his own sphere. Here we may cite another source of evidence. In many instances, excavations of the graves of ancient kings have revealed that when a king was buried, members of his court were put to death to accompany him to the next world and that the persons sacrificed submitted voluntarily and with equanimity to their fate.

This is difficult for the modern mind to grasp. We think of human sacrifice as brutal murder, yet to primitive people it may not have had this implication. In the first place, the prize we have won, the discovery of self and the development of self-determination, implies also a loss, the loss of a sense of the reality of the divine, and of divinity operating through appointed leaders. When this sense operated, as it must have in earlier times—for we have from Egyptian, American Indian, and in fact all early traditions, that their civilizations were inaugurated by gods who taught them to plant by the seasons, to spin silk, to play flutes, etc.— when such implicit trust in their leaders operated, there was no conflict of interests between the individual self and the group self because there was no consciousness of an individual self. If the leader died, the self, being a creature of the leader, died too.

We may go even further and speculate on an actual consciousness of deathlessness, a sense of immortality in primitive man at the collective stage. While this sense, again, is foreign to the modern way of thinking, we must keep in mind that primitive people, by their continual consultation of the gods through oracles and sacrifice, and their dependence on such consultations for the conduct of affairs, must have lived with a sense of the continual presence of the divine. They would thus have an altogether different attitude toward self-sacrifice on such important occasions as the death of a king.

We advance this last point not only because it helps make sense of the facts of primitive life, but because it lends confirmation to the suggestion we advanced earlier, that the first two stages of process are indestructible, that is, immortal. Finite existence in time, like finite existence in space, can come about only at the third and fourth levels, when forms and formed substance come into existence. Such *forms* are destructible, while the basic ingredients of which they are composed, substance and function, are not destructible.

Gautama Buddha touched on this subject in his last words, "Brethren, I remind you this (death) is the end of all compounded things."*

Third substage

Before man is capable of discovering and handling laws by which things are combined (the laws of science and society), he must be conscious

---

* In Hindu religion the doctrine of Karma is one of the important primary principles. Can we not read this principle, which states the indestructibility of desire energy, as a generalization of the conservation of energy?

of himself as different from other persons. This is something of which an
animal is not capable. I recall an incident with our dogs. The recently
acquired puppy had perpetrated a mistake on the carpet. He was being
punished, yet he showed no evidence of guilt or sorrow. He simply
wagged his tail in the most genial fashion. The older dog knew what it
was about; he cringed and looked very miserable indeed. He was
conscious of guilt, undoubtedly, but could not distinguish the puppy's
mistake from one of his own.

Man learns the laws of things by observing the results of what he does
to things. Even as witness to the action of a play, he must identify
himself with one of the characters if he is to be moved by it, even to feel
the tragedy or be interested in the plot. The ego, the mouth through
which self draws the substance of experience on which knowledge is
built, has its roots in self-identification.

We would be inclined to interpret Greek civilization as third-
substage, a time when people began to question authority and think for
themselves. Indeed, the rational mind, which demands a reason for
everything, is itself a typically third-stage phenomenon. The third-stage
power of identity being acquired by creating a center of its own breaks
away from totality and becomes a world to itself. Its world is small, it is
but a fragment of the whole, but it is rational and self-governed.

Of course, we know little of early civilizations, but from what we do
know, it is clear that the modern mind emerged with the Greeks. The
Greeks pioneered in the making of abstract concepts and in
distinguishing concepts from physical objects. They were the first to
question the authority of the gods. They asked questions and learned
to reason. Man became responsible for his mistakes. Prior to this,
misfortunes were the punishment inflicted by the gods. (On the whole,
the earlier reliance on authority worked pretty well, as the Egyptian
kingdom survived for four thousand years.) By their bringing gods
down to a human level, the Greeks made man self-determined. Their
questions led to science.

The mistakes of self-conscious behavior bring on the fourth substage,
the necessity for correcting first guesses by reference to fact. But in the
third we are concerned with the *origin* of error, which requires
consciousness, a ring of light that illuminates the self and its immediate
vicinity.

In the growth of the individual we can detect this third substage as the self's break with authority, first with that of the mother and father, and later with that of its own group. This is what psychologists call individualism, the discovery of identity. It is the beginning of self-determination. It is only because of such self-determination that the self can now profit by its interaction with others and learn the lessons possible in the fourth substage.

A difficulty is that we cannot tell to what degree Greek life also partook of the fourth stage. There were laws and disputes between people, to be sure. But there was not that all-out belief in objectivity, in research, that now prevails. The very widespread dependence on science which distinguishes the present time from all others brings with it a special emphasis on the secular—the mundane—that makes it seem as though the present were lost to all else.

However, our task is not the assignment of civilizations to different substages, but the describing and illustrating of the substages of the individual's development, because in the seventh kingdom, especially, it is the *individual* that is evolving.

Fourth substage

This is a more advanced state than the collectivity of substage two, in which the units are undifferentiated and without identity. It is also a state that presupposes the self-determination and self-consciousness of substage three. But it has not yet mastered the principle of combination as has substage five, for modern man is at best still immersed in the problem of finding his own boundaries, of meeting the consequences of his own acts, of learning the law of cause and effect.

This is the fourth stage in any process, the stage to which we assigned the power of combination, but by implication it is also separation, and its task is to learn the law so that it may then use this law and move on to the fifth, with its power of growth and generation of seed.

The emphasis on science, the knowledge of natural law, which characterizes modern times is a particularly appropriate indication that modern man is at the fourth substage. Moreover, since science does not limit itself to practical applications, but puts it down as a credo that laws are omnipresent and rule all phenomena, we may conclude that man's dependence on religion caused him to translate science into a belief system.

Science, then, is one symptom by which we diagnose modern man as fourth-substage. But there are other symptoms. The task of *combination* is a fundamental preoccupation of modern man. For one thing, he is occupied with combining material substances in construction of buildings, ships, railroads, and the host of devices which encrust our civilization. Combinations of people are also involved: marriages, partnerships, corporations, the combination of capital and labor, of man and machine, of money and goods. Such combinations undergo incessant activity, with constant struggle to achieve working solutions: man against man, man against machine, always searching out a *law*. When the law is discovered and established, focus shifts and the struggle moves to a new issue. At one time it is property rights, at another the right to vote, now the right to strike, now the right to work. But always there is an increasing body of law—social law, moral law, corporation law, scientific law—a striving to have everything be *determined* by law, leaving nothing arbitrary and lawless. (Eighty percent of American presidents have been lawyers.)

The automobile, which gives the owner independence but makes him responsible for his own and others' safety, is a major contribution to the fourth-stage nature of contemporary life, and I would hazard a guess that the evolution of man has been definitely speeded up by the control learned through its use.

There is one further symptom, one that currently has become acute, that places modern man not only in the fourth substage, but in the middle of it, at that critical point we call the "turn." This symptom is that the very devices that man has invented for the control of nature now appear to have gotten out of hand and pose a threat to his survival more serious than those of nature. Pollution of the environment, the atmosphere, and the ocean, overpopulation, the possibility of atomic warfare, are threats that are a consequence of man's own actions and have created new problems he must solve. The answers, whatever they may be, require that he take action of a different kind from the actions that have heretofore characterized the "progress" of civilization. To a great extent, what he has already learned with regard to putting the forces of nature to his service has been by trial and error, by observing cause and effect, but each step in this conquest has been followed by an extension into a wider sphere, which in turn brings new challenges.

This outward expansion has only recently run into trouble, which has developed because the *space available is not unlimited*. The planet is a finite globe; its surface curves back into itself. The poisons we throw away return. Even the improvements have unfortunate side effects: medication preserves the unfit, fertilizers disturb the ecology, insecticides kill the birds, spending for prosperity creates inflation. It becomes evident that the laws of addition invert when applied on a scale which exceeds the limits of the available space, and that man must exercise vision and take into account his limited ability to assume the responsibilities that nature has hitherto borne and whose extent and subtlety we are only beginning to appreciate.

The word that covers this new awareness is *self-limitation*. It is a word that suggests coldness, confinement, restraint, but, by the paradoxical nature of the vital principle, its consequence is inner growth. Our paradigm is the life cycle of the obelia (see p. 121). This animal having attained a form and the ability to move about in its blastular, or third, stage, *fastens itself to the ocean floor* at its fourth stage. This self-limitation, which surrenders the freedom of random exploration, becomes the basis for its true expansion, which occurs through its growth into a large plant-like organism (fifth stage). This analogy may seem arbitrary, but it will be recalled that the obelia is the example chosen by biologists to represent the life cycle of *all* organisms,* and that its life cycle correlates step for step with that of the human embryo, in which the fertilized egg, having become a free-swimming multicellular organism (blastocyst), attaches itself to the wall of the womb, then to grow into the form which it has at birth, when it begins to function as an animal (sixth stage).

Whatever the degree to which we draw on the obelia (which we offer as illustrative), and whether we are guided by empirical examples or deductive principles, we must see the promise of a future for man as dependent on an inward, self-generated reformation, like the drunkard who *himself* decides what are his own best interests, or the prodigal son

---

* Even the much simpler alga, a primitive plant, has a stage in its life cycle when what is called the holdfast cell attaches to the ocean floor, then to grow into the multicellular plant (seaweed).

who returns. This inner resolve is not done for man either by nature or by the state. It is, rather, a point in man's own drama to which all nature is spectator, in which man is the lonely protagonist.

What makes this decision most difficult is that man today has lost faith in his spiritual origins, a conviction brought about by the very nature of the fourth stage with its belief in determinism and in objective laws which supposedly are alone able to cause effects. Thus we find man, having created science—the tool of all tools, whose every discovery is an extension of man's will—denying in the name of science the very power by which he created science.

Fifth substage

Because the scheme we are constructing deals with people—or rather with the growth or evolution of the monad (and we may just as well get used to this term)—the fifth substage is quite clear because we can assign to it outstanding people—Galileo, Goethe, Napoleon, Mozart, Shakespeare, Beethoven—the great of history. But the selection of such persons should be based on objective considerations.

In the interests of objectivity, let us see what theoretical criterion we can establish for the fifth substage of this kingdom. Recalling that the fourth substage is "learning the law," we could expect the fifth to begin *when the law was learned;* when the entity could, like the cell which begins life, manufacture out of itself ingredients it needs to organize something much larger than it starts with. This is not just learning the law; it is mastery of the law. The implication is that persons at the fifth substage would show outstanding competence which might even be evident early in life.

There are many examples of the early appearance of exceptional ability in the great. We have Mozart writing symphonies at the age of seven. Newton, not considered precocious, as a small boy made mechanical toys, including a mill propelled by a mouse that ground flour, and a clock that kept time. Pascal at the age of sixteen proved an important theorem upon which the whole of projective geometry depends. Galois, before he died at the age of twenty-one, had made a contribution which anticipated fifty years of mathematical progress.

The emergence of exceptional ability at an early age is not necessary
to genius, nor is it a criterion for the fifth substage. The examples
of early emergence are cited rather to emphasize the nature of the
problem. How are we to account for exceptional abilities except as a
heritage of the evolution of the monad and not as genetic inheritance
or environmental training? Bell,* describing the life of the
mathematician Gauss, tells of an incident that occurred before the age
of three. The father was making out the weekly payroll for workers
under his charge. Coming to the end of his long computations, he was
startled to hear the young boy pipe up, "Father, the reckoning is wrong;
it should be. . . ." A check showed the boy correct. The abilities of
Gauss were phenomenal throughout his life, but like other infant
prodigies, they did not start at zero; they were manifest as soon as he
could talk.

The problem of human evolution: genius

The problem that we are dealing with throughout the seventh kingdom,
and even throughout the grid, presents itself with greatest impact here at
this fifth substage, when we come to consider the great creators, the
great leaders. How did this greatness come about? To attribute this
greatness—whether to the infinite capacity for taking pains (as did
Carlyle in his *Biography of Frederick the Great*), to inspired genius, to
a determination to overcome difficulties, or to a surpassing ability to
influence people—is not to explain it.

But the fact that it is great men, the exceptions to the general run, that
bring the problem into focus is itself significant. Is it possible that we are
overlooking something to which ordinary life is heir? At the risk of
possible repetition, let me remined the reader of what he himself
consists. There are:

Some 10^{12} (1,000,000,000,000) cells in the body.
Some 10^9 (1,000,000,000) molecules in the cell.
Some 10^6 (1,000,000) atoms in the average molecule. (DNA has
   billions of atoms, others thousands, and some only a few.)

* Bell, E. T. *Men of Mathematics*. New York: Simon & Schuster, 1937.

Every one of us is such a hierarchy, an organization of some 10^{27} atoms all carrying out their appointed tasks, in order that we may read a book, hoe a garden, drive a car.*

Seen in this light, the human being is already fantastic in the complexity of his organization. Yet we seem to need the example of genius to force us to raise questions about how it can be possible for an entity to be composing music at four, and writing symphonies at seven.

In the case of the genius—the mathematical, the artistic prodigy— we cannot trace genius to ancestry. The ancestry of great men is in many cases quite ordinary and in no case accounts for genius. Again, great men do not have remarkable children. In short, we must find something other than ancestry to account for greatness.

As we said before, the group soul concept, which we have evoked to account for the elaborate instincts of animals, does not apply to man. Here we are dealing with the individual, and we can see no way in which individual competence or greatness can emerge without many lifetimes of development. Once we recognize that greatness requires such development, it becomes evident that even normal human capabilities require it to some extent. This, in fact, is the implication of all stages and substages of the grid: in each we can see the need for the inheritance of the accumulated abilities and powers that have gone before.

The grid is in this sense a table of debits and assets, of investments and expenditures, of accounts payable and accounts receivable, of the entities of nature. It is a scheme of accounting because it endeavors to trace all acquisitions to their source and hence to reduce undefined terms to a minimum. It is the principle of least action** applied to the process of evolution. In other words, it shows the shortest path by which the complex can evolve from the simple. To deny a heritage to the complex would be to assert that the complex emerges spontaneously.

We have insisted on the necessity of a continuously operating principle of spontaneity at the core of life, but we have not invoked this principle to supplant cause and effect. If it may seem otherwise, it is

* This argument applies also to animals, but that does not weaken its force, because man is even more complex.

** The principle of least action states that natural process occurs with a minimum of "action" (action is change of inertia with respect to time) (see pp. 16–17).

largely because other accounts have overlooked the difficulties, have ignored rather than explained the problem.

To continue, at the fifth substage we are to expect the emergence of an ability corresponding to that of plants (fifth stage) not only to grow to large stature, but to create seeds which have a life of their own. The power of reproduction lies as much in the letting go of authority (permitting uncertainty) as in the transmission of a pattern (enforcing certainty).

This becomes the criterion for fifth-stage persons, whether writers, painters, composers, mathematicians, physicists, statesmen, or political leaders: *if their works,* like the seeds of plants, *have a life of their own and outlive their creators, they could qualify as fifth-stage.* Contemporary success, since it arises from the person himself, is not enough.

The difficulty that confronts us in the seventh kingdom is that since we ourselves are in it, we do not have, as we do in the first six, the benefit of a perspective which permits viewing its creatures objectively; nor do we have the benefit of scientific classifications. We have not only to define the classifications, but to assign entities to them.

We cannot use the classifications ordinarily employed, as, for instance, authors, statesmen, composers, scientists, etc. These are horizontal classifications which need have nothing to do with competence *per se*. We must view all professions as equally qualified. Again, we must discount contemporary figures, and let time be the judge of the factor we have made crucial, namely, vitality of works. Let us therefore select a few typical examples from history and from various callings.

> *Painters*: A number of great Renaissance painters, for example, Giotto, Leonardo, Raphael, Tintoretto, Titian, and many more; Rembrandt, Rubens, Vandyke; also recent painters, the Impressionists. (Modern painters are omitted because we still do not have the testimony of time.)
>
> *Composers*: Bach, Beethoven, Brahms, Chopin, Haydn, Mozart, Wagner.
>
> *Statesmen*: Bismarck, Alexander the Great, Elizabeth I, Garibaldi, Napoleon, Washington. (Note that each of these has supplied the initial impulse—the seed—for a new nation or a new regime.)

Authors: Blake, Dickens, Goethe, Shakespeare, Molière, Yeats.
Scientists: Euclid, Clerk-Maxwell, Galileo, Kepler, Leibniz, Newton.
(Again note the contribution to the later development of science:
these men were creators, founders of new concepts.)

Such a list is made only to supply examples of what we mean by persons who have produced the seed, the creative impetus, for new developments, for what could be called cultural advances.

Individual ego still the vehicle

We should now make it unmistakably clear that our position with regard to the evolution of man does not imply that man is, as it were, a cell in the social body. Man in our view is not at this stage evolving toward a collective superorganism. Teilhard de Chardin has been so interpreted, and there is a general disposition to envisage that which is beyond man as a kind of social or racial entity of which individual men are cells. That is not what our theory of process indicates. What we believe to be implied by the theory is that the seventh stage involves evolution *through individuals,* at least up to the end of the fifth substage. We believe that great creative individuals provide concrete evidence of the fifth substage of the dominion kingdom. The fact that some men have achieved this level of competence or power is evidence that it is an evolutionary stage and hence to be traveled by *all* monads in their development.

Even with the aid of the theory we have set up, we find it difficult to express what Genesis states quite simply:

For God doth know that in the day ye eat thereof then your eyes shall be opened and ye shall be as gods, knowing good from evil.

"Your eyes shall be opened"—this is what we have referred to as the "turn." We have already stressed the importance of the turn as the basis for the higher stages of process. It is the point at which the entity, having expended its initial endowment of free energy, "wises up to itself" and starts to control itself. An analogous control of energy made possible the complex molecules of cellular life and hence the plants.

We could liken the power of plants to convert sunlight into order to the power of genius to convert the divine outpouring into works. Nor need we question this "outpouring," for whether we call it divine or natural, we are aware of its presence, though it is only the genius who is able to capture it in works.

If this argument fails to convince, we can still point out the logical necessity for a stage of competence in the evolution of the human monad that corresponds to the power of plants to grow and reproduce. Such competence is equivalent to the creative ability of the great ones of history. It is often insisted that spiritual growth can forego this worldly competence, a position to which religious persons incline, but I would urge that however noble the surrender of personality, like the sacrifice of worldly goods, this sacrifice can only be authentic after personality has been fully achieved.

In other words, I insist that merging into a superorganism, if it occurs at all, can occur only after the individuation has been completed, at the end of the fifth substage. For just as the third substage sees the separation of the self from the group, or the start of individuation, so the fifth substage would see the end of individuation and the reunion of the self with the group, bringing to it the value learned from the experience it gained as an individual.

Such are our reasons for rejecting the interpretation of the stage beyond present man as a superorganism, related to its component cells as an army to its soldiers. Even this image misconstrues the nature of the army, since the army is only as good as its component members, be they generals, officers, or soldiers. Organizations like the army are in this view training ground for the self, which must reach an overall competence in its sphere equivalent to the general's in his before its graduation from the fifth.

So we must realize that the higher substages of evolution (beyond the fifth) can occur only after the fifth has accomplished its particular task of self-growth. We can, however, view the propagation of the seed, which is a kind of surrender of self, as the beginning of dedication to something far greater than the individual self.

So we assign to the fifth substage a considerable task. It sees the development of superhuman competence and the creation of works, either works of art, new concepts or creations of nations, which outlive their creator. While such acts are essential in the development of civilization, and may thus lead to the notion that civilization has a kind of life of its own, we reiterate that it is not an end in itself. It is rather the theatre or playing field in which the monad finds facilities for practice.

Kundalini, a fifth-stage phenomenon

We have so far dealt with rather intangible aspects of man's evolution by using the criterion of works, of effects rather than causes. We can describe the plays of Shakespeare as having a life of their own, but we know nothing of the genius who produced them. This is to be expected, for our method does not pretend to reduce the powers to something other than, or simpler than, themselves. Thus we must continuously refer to examples in our account, whose justification rests in sorting out the unknowns and reducing their number, rather than explaining them.

So cautioned, let us consider a piece of evidence described by the Hindu tradition. This is the serpent power* or Kundalini of Yoga teachings.

Kundalini is a "force"** which, according to advanced practitioners of Yoga, is evoked by meditation. This force is said to arise at the base of the spine and to flow up through the spinal column, linking the various nerve centers. These centers are the Root Chakra, Spleen Chakra, Navel Chakra,*** Heart Chakra, Throat Chakra, Brow Chakra, and Crown Chakra. The chakras are not recognized as such in Western science, but correspond to the ganglia or centers of the autonomic nervous system and possibly to the endocrine glands (located approximately in the same areas). However this may be, the Kundalini force would constitute a *linking* up of a series of centers which control the human organism.

Assuming that the power of dominion, however intangible it may be, must have *some* physical or quasi-physical embodiment, and recalling also that the average man has little or no conscious control over autonomic functions, whether these be sexual energies, digestive functions, circulation, etc., nor over such glandular functions as control of growth and other regulations of the body, we might expect there to be some linking together of these subhierarchies in the evolution of the dominion principle.

Kundalini meets the requirements of such a linking-up, and if awakened, not prematurely, and under adequate self-control, it

---

* Woodroofe, Sir John George (Arthur Avalon). *The Serpent Power*. 6th ed. Madras: Ganesh & Co., Ltd., 1958.

** *Shakti*, in the original writings.

*** Below the heart, chakras are given several variant listings, which do not affect the present argument.

subordinates the energies of the sexual and other centers to the self's command. Such a linking is distinctly fifth-stage, which we have often likened to a chain, the chain of cells of which vegetation basically consists, the chain of generations in propagation, the chain of command in hierarchy. We also have the fifth substage of other kingdoms, the polymers (chain molecules), calamites (segmented vascular tissue), and metamers (animals composed of a chain of segments).

Therefore we believe that Kundalini fills the chain role that we could expect at the fifth substage of the seventh kingdom or domain. We could further suppose that these great men and women, with their superhuman endowment of creative force under conscious control, already employ this "force" by making it available to the central command of self.

Transcendence of the personality

We have described the fifth substage as beginning when the monad achieves the ability to create works that live, an ability analogous to that of plants to create seed that has its own life. How are we to mark its end?

Since powers are cumulative, we would not expect any termination of this ability. So we must look for a new power—which transcends by changing or *transforming*. But we need other guidelines. What is the task of the fifth?

Here it is pertinent to refer to the levels. The stages from 3 to 5 carry the monad through the two lower levels.

1 Purpose   7 Goal

2 Substance   6 Mobility

3 Form   5 Organization

4 Objects

Stage 3 introduces the possibility of *form*. For the monad, this is an *ego*, or an individual vehicle, a limitation necessary to its early evolution, but this limitation later becomes a handicap. At the third stage, the centeredness it provides makes knowing possible. Genesis refers to this as eating the fruit of the tree of knowledge of good and evil, and affirms that the use of this diet for a period of time will make men "wise as gods, knowing good from evil."

We have indicated that the genius has mastered the law—how to do things, how to create, to paint, to write, to compose, to govern, to act—in short, power. But the knowledge of good and evil implies more. It implies not letting this competence become an end in itself—not becoming corrupted by power. We had implications of this in the growth principle, which must learn to create seed which drains off this power—a power which would, without this outlet, destroy the vehicle.

This inability to handle absolute power is due to the finite limits of the ego, the very property that made learning possible. So if evolution is to go further, it has to transcend the ego—the personality. I suspect this would eventually lead to no longer needing a physical body, since even without self-aggrandizement, the energy involved would put more strain on a body than it could take, just as too much voltage will blow out a light bulb.

Transcending personality is thus the task of the fifth stage, perhaps its midpoint, and its upper limit would be that point in the monad's further evolution when it learned all it needed from a selfless dedication to great tasks *without* the inducement of personal aggrandizement. In a survey of human history, it is not hard to distinguish between the more conspicuous personalities—Byron, Queen Elizabeth, Paganini, etc.—and the less conspicuous men of even greater power—Goethe, Frederick the Great, Bacon, and others such as the remarkable artists of the Renaissance: Titian, Giorgione, Leonardo—of whom, compared with those with charisma, we know very little besides their works. Certainly there are lesser and greater geniuses, and if giving up personality is the task of the fifth substage, we can make this sacrifice its pivot.

Sidelights on genius

Modern life suffers from its efforts to pull everything down to its own level, to debunk—and it is worthwhile, if only for our own entertainment, to take a fresh look at genius. George Washington may not have thrown a shilling across the Potomac, but Willoughby, a conscientious compiler of records of athletes,* states that one of the first records in the running broad jump made in the United States was one of twenty-

* Willoughby, David P. *The Super Athletes*. New York: A. S. Barnes & Co., 1970.

three feet, said to have been made by George Washington when he was eighteen, and that this record stood for over a hundred years. Washington was also a first-class wrestler, as was Abe Lincoln.

Willoughby also states of Leonardo da Vinci that "one of his feats, with which he used to astonish his visitors, was to jump upside down to the ceiling and kick the bells of a glass chandelier." Douglas Fairbanks, Sr., according to Willoughby again, could, in one step, jump with both feet onto a platform five feet above the floor. In my own reading of the lives of some who must by their other abilities have been high in the genius range, I found with both Benvenuto Cellini and Cheiro (the palmist) that people who had aroused their ire fell down before them without having been touched. Cellini in his autobiography relates a number of incidents that imply superhuman physical abilities which would be unbelievable were it not that his works of art equally surpass the limits of what one can believe.

I do not mean that athletic prowess has anything to do with genius. It is their works that earn for them this designation, but I am trying by these references to physical prowess to direct attention back to the source, the monad—to reawaken a sense of the marvelous strength of a monad.

The point, I believe, is that the works of genius, however great, are inert and can be adjusted to, whereas the monad behind the works is vital and creates each time anew. It is thus more elusive, more universal, more to be admired than are its works—and, one should realize, that the monad, whatever it is, not only can paint, invent, etc., but it enlivens and animates the body.

Another sidelight that brings out this point is the tenacity for life exhibited by some historical figures. Sir Thomas Overbury, murdered by his political enemies, withstood enough poison to kill twenty men. Rasputin, the "Mad Monk of Russia," resisted poison, knife wounds, and gunshot enough to kill an ordinary person many times over. Swami Rama, the practitioner of Yoga tested at the Menninger Foundation, not only stopped his heart at will, but ate potassium cyanide to demonstrate the power of mind over body. These feats, which in isolation are mere wonders, acquire valuable significance in the present context, for together with works of art, power to influence others, etc., they give evidence of the increasing evolutionary stature of the monad. This will be even more important when we come to the sixth substage.

End of the fifth substage: indications of what is to come

How are we to characterize the end of the fifth substage? We can at least establish its lower limit as that point in the evolution of the monad when it had acquired all that incarnation as an ego required; when, to anticipate our next chapter, it had accomplished the "labors of Hercules"; when it had eaten of the fruit of the tree of knowledge of good and evil and discovered the tree of life. Substage six is beyond personality.

The end of the fifth is a point where we can begin to fit moral principles into place along with evolutionary ones. What Christian tradition refers to as the "fall of man" is the descent of the life spark, itself of divine origin, into a *mortal* existence. Why? So that it might know good from evil, or in terms of process, learn the law and use it.

How may it do so? It must first acquire an ego, a center from which to initiate acts and from which to view the consequence of these acts. Thus, and only thus, can the monad learn, for the moral consequence of an act has significance only to the one who initiates it. One must "own" a thing to learn responsibility.

If A intentionally injures B, B suffers, but the moral consequences rebound or flow back on A, creating the need to right the wrong, an adjustment that may require time, or even another lifetime (the basis for the Hindu doctrine of Karma).

Thus Karma, which is the law of cause and effect manifesting at the psychic level, leads to the knowledge of good and evil, and becomes the basis for the monad's growth, its conquest of matter. At the end of the fifth substage, it has learned this lesson. It no longer needs the center it required for self-determination and the experience of the consequences of its acts.* We can assume that it can now become what Joan Grant calls the integral, the synthesis of all its previous existences.**

* This may give a clue to the loss by Horus of the eye, that curious detail in the battle of Horus with Set. As will be recounted in the next chapter, the god Set traps Osiris in the jewel-encrusted casket (the compulsive desire that precipitates the fall). The casket floats down the Nile (the fall). Osiris is dismembered and reborn as Horus, who eventually conquers Set. But Horus is deprived of his eye. My hunch is that this may be a reference to *no longer having an ego.*

** Joan Grant, author of *Winged Pharaoh, The Eye of Horus,* and other books on her past lifetimes, uses this term to describe the higher or "total" self.

What the monad has achieved through its long evolution is not lost. If we use the difference between animal and vegetable as an analogy of the difference between the sixth and the fifth substages, the sixth is concerned with *animation*. Form as such is transcended and can now be freely varied and used. Emphasis shifts to the dynamic or energy aspect.

The battle here is to conquer, or overcome the compulsive nature of the desire that at the second stage precipitated the fall. (Desire is dynamic as opposed to form, which is static.) But this phrasing is scarcely adequate for an appreciation of the sixth substage. Let us look deeper.

Beyond genius

A clue is the quality of the work of the great painters. I will have to run the risk of making value judgments about art. My premise is that some of the Renaissance masters—Titian, Correggio, Veronese, Leonardo, and other very great painters—exhibit skill of a higher caliber than, say, that of the Impressionists, whose paintings may be judged more beautiful, but which still do not show the degree of technical mastery that is evident in the earlier painters, and this ability is a clue to the sixth substage.

What I refer to especially is texture, the extraordinary living-ness of the flesh in Titian, or the hypnotic expressions in Leonardo's paintings. What this suggests to me is the observation of substance (in nature) and its execution in paint, as compared with painters who are dealing with finite composition. The achievement of the Impressionists was quite sensory, but it is still more a question of composition analogous to musical composition. It deals with only a finite number of elements, like the notes of the musical scale, whereas Titian deals with a continuum, and the emphasis shifts to vitality itself. We feel that if we look through the microscope at the flesh of Titian, we would see cells, nor is there sacrifice of large-scale vitality. It is, rather, that the painter has a greater range. In the case of Leonardo, this range exceeds limits of paint. In a sense, if his paintings are not beautiful, it is because he is trying to force more from paint than it can yield.

This search for superhuman goals is not religious or conceptual as in the painting of El Greco or Blake, which I admit is ecstatic, passionate.

It is a discontent with drama and with finiteness, with anything artificial, whether it be an artifice of imagination or an artifice of intellect. It is an effort to plumb so deep that the very molecules resonate.

This, I think, is the shift in emphasis that points to the sixth substage, because, like the examples available from the sixth column of the grid, it is a shift from organized form to animated fluidity, from "growth" to "mobility." Note too the attention to *substance* in these great painters.

Fascinating how we go from intellect to emotion, from nonphysical to physical, on our way up to spirit; one would have thought intellect nearer to spirit, but it is not nearer in the steps of development. Substance is higher than form. Castaneda's quest to convert the threatening animals into allies is this step.

The magical: transcendence of reason

Now in calling attention to this emphasis on substance in the great painters and in psychometry, I do not know for certain whether this is sixth-substage. I am, rather, trying to indicate this power (which the highest type of genius exhibits) as a clue to sixth-substage, a clue which points in the direction of what now becomes a theoretical possibility, the *magical*.

Here we are in direct conflict with reason, which would rule out magic at any level on the grounds that the laws of matter forbid it, but we have grounds for predicting magic here. Magic is the control or manipulation of illusion, and is expected here because sixth is an inversion of second, which was being caught up in illusion. Magic is also suggested by the two degrees of freedom at the second level, which allows content to change from one form to another.

Form, we repeat, is here transcended and subject to manipulation by motive. Motive is the only constraint. This is the "magician" formula. The magician of fairy tales transforms himself into a seed; his rival transforms himself into a hen; then the first one into a fox, etc. We are saved from complete unconnectedness by the fact that the chase goes on. The participants are still rivals. Such duality, of course, is characteristic of second-level.

We have assigned Christ and Buddha to this state in the grid. Other historical figures, such as Zoroaster, Dionysus, Krishna, and Orpheus, might be included. Such assignment is not intended as an evaluation of these beings as much as it is a way of giving a place in the scheme of things to the existences that legend and religion have described as gods. Almost all early civilizations credit their origins to gods who came down and taught man the cultivation of the silkworm, how to weave, the cultivation of corn, the arts of music, etc. Beings of this order, according to the evolutionary scheme we are setting forth, must have existed and must still exist, modern thought to the contrary.

We might call attention to the Biblical account of the genealogy of the ancestors of Noah, Genesis, chapter 5. With the exception of Enoch (whom God took at the age of 365 years), Noah's ancestors lived an average of 900 years, including his grandfather, Methuselah (969 years).

Genesis, chapter 6, verse 4, follows with the provocative passage:

> There were giants in the earth in those days and also after that when the sons of God came unto the daughters of men and they bore children to them, the same became mighty men which were of old, men of renown.

The modern mind, having no scientific evidence for persons living 900 years, or for "sons of God," would presumably dismiss such accounts as the distortion of myth.* When, however, we have a theory that anticipates transcendence of mortality, accounts describing it are seen in a different light. They may refer to what we are calling sixth-substage beings, whom we would expect to have conquered mortality.

In this context, the Babylonian *King List*, also dealing with the period before the Flood, gives still longer periods for the reigns of the early kings.

In any case, rather than reject all such references as absurd, if we purport to be scientific we should try to find ways to overcome our mental blocks in this direction, not to become gullible, but to constructively speculate about what lies beyond our immediate environment and our present status.

To illustrate the manner in which rationality tries to set limits on things, to put a roof over our heads, let us consider astronomy.

* Or of time scale, as some scholars have surmised. However, if we change years to months, which would give Methuselah the age of 73, Enoch would have been 5 years old when he begat Methuselah.

The astronomers have been repeatedly forced to expand their concepts of the universe. This enforced conceptual growth began with the telescope, through which Galileo purported to see markings on the moon, moons revolving around Jupiter, and such.

Many of his contemporaries refused to look. But as the years passed, others made telescopes. Newton solved the problem of planetary motion, and the heliocentric concept of a solar universe came to stay.

Instrumental techniques followed, and it became possible to measure stellar distances which exceeded that of the sun's distance by a factor of a million or more. It became evident that the stars were suns, inconceivably remote by earth standards of distance.

Then came the discovery of cloudy patches in the sky, which were first believed to be clouds of gas and were called nebulae. These, as the reader of course knows, turned out to be distant universes, billions of them.

Now I am not subjecting the reader to this recitation to ask him to engage in astronomical one-upmanship, since the very remoteness of the subject leads to excessive speculation. My point is that reason, with its compulsion to set limits, tends to block out truth.

The new-found facts formed a mind-expanding sequence—the solar universe, the galactic universe, the universe of galaxies—which, like the continued use of a drug, caused a sort of immunity to magnitude. Astronomers speak of distance in terms of light years, or exponents of 10, and cosmologists put the universe in its place and console their egos by what is called the cosmological hypothesis: the universe, except for local irregularities, is the same for all observers. A solar system is thus "a local irregularity." So too is a galaxy. This insistence on symmetry has a medieval flavor. It treats the universe as if it were an ideal gas or the grains of sand in a desert. Once again, intellect attempted to close the wound in the skin of the self-conscious ego. As in the Beatles' song, we are continually fixing holes in the roof to keep our minds from wandering.

Then came the stellar probes. Automated rockets operating outside the earth's atmosphere detected ultraviolet light sources* that put out

---

* Since ultraviolet light cannot penetrate the earth's atmosphere, the fantastic energy output of these objects had not been apparent to observers on the earth's surface.

energy at a rate billions of times that of the sun. Such existences not only taxed beyond credibility the concept of "local irregularities," but they violated well-established theories that stars could not exceed the size of the sun by a factor of more than 100. Other recent findings, pulsars, quasars, etc., indicate the utter inadequacy of existing theories, but I have no doubt the boys will soon catch up.

I will not carry my brief account further, but cosmology remains awe-inspiring despite efforts to wrap it up, and the moral is: don't trust the limited boundaries which the rational mind uses to protect itself. Don't permit statistical law to give the illusion that there is nothing here but us chickens. In other words, don't conceal evidence, even the evidence for the potential divinity of man. According to Psalm 82 (chapter 6), God said:

> Ye are gods and all of you are children of the most high.

Postscript

I am sometimes asked why I do not include in my theory the evolution of stars and galaxies. The grandeur of stellar processes, of galactic evolution, is indeed wonderful, mind-boggling, the discoveries of astronomers fascinating, but I cannot see that the subject has relevance except as having a correspondence, on a much larger scale, to the planetary evolution we have discussed. Other than that, we have no evidence (such as we had from the kingdoms of nature) that could carry our study into galactic dimensions. Through telescopes the stars can be discovered as points of light, by spectroscope analysis and other methods, temperature, velocity, density, magnitude, etc., of stellar bodies can be established, but these data give no clue as to the life, the consciousness of the stars, and as we noted above, they constitute a one-sided and possibly misleading kind of evidence. (Why did it not give us evidence earlier of stars billions of times more powerful than our sun?)

It would be better just to look at the stars on a clear night and be invigorated by the sense of wonder.

XIV | Beyond man

We are at last reaching the end of our study, and we may well try to marshal the many considerations that have been taken up around a unifying and central theme. We stated in Chapter IV that process begins with a particular, though undefined, dynamic that we called "purpose," which goes out into matter to obtain means to effect this purpose and finally attains its end: it actualizes the purpose that first sent it forth.

Our central theme is that a self, or a universe, is of the same nature. It starts and ends with undefined purpose, unrestricted choice. We may variously describe this as free choice, spontaneous causation, randomicity, but in any case it is not committed to objective form, terminal state, or to any local or temporal manifestation. To illustrate the transcendency of form which characterizes purpose, think of the purpose "to travel," then of the variety of forms used for traveling: horse and carriage, automobile, airplane, space rocket, and finally, perhaps, teleportation.

Let us not forget the difficulty which our hypothesis encounters. This is that the essential nature of the first principle is undefinable; *it is knowable only through its effects*. It is similar to that remarkable experiment familiar to physicists called the Wilson cloud chamber experiment, in which the path of an invisible particle (which may be a proton, an electron, or other nuclear particle) leaves a vapor trail which is photographed as a bright line. We see this bright line, but cannot see its cause (the particle is so small that it cannot possibly reflect a light wave, its dimensions being less than 1/1000 of the wavelength of visible light).

207

This first cause, or as Aristotle calls it, *final cause*, is both beginning and end, the first and last stage of process. It is both the light kingdom and the dominion kingdom. Its nature is complete freedom. It has but one law—the law of hierarchy. It is not determinable; it has no form; it is not objective. To attempt to describe it, we call it projective or dynamic, and to distinguish it from the generality of the other formless dynamic (energy or raw substance), we call it particular. All of this wordiness could be avoided by calling it "spirit," which we can now do with some impunity, for we have erected a scaffolding to enclose its formlessness and hence are not open to the charge of vagueness.

This word "spirit" is a very good word, despite the fact that it is presently out of fashion. It is rather similar to "sublimation" in that it too refers to a physical operation—spirit being that which is distilled from wine or other admixture with water.* The point is that it is not so much what word we use that matters as the implication with which we endow these words. By the phrase "merely a sublimation of animal drives," we may imply that man is "no more than" animal; but when we do this, as we said in Chapter I, we are also abolishing the distinctions between other kingdoms. We are implying that the animal is merely a collection of chemicals, and the chemicals are merely atoms. Carried out completely, this course would take us to electrons and protons, which we find are not things at all; they are "probability fogs." So we are left with nothing whatever except the metaphysical concept that all manifestation is illusion. There is nothing wrong with this concept either. The point is, the rather irresponsible statement that man is merely the "sublimation of animal drives" implies the removal of *all* distinctions that exist between kingdoms, and we need these distinctions for an account of evolution.

The distinctions between kingdoms, as we have endeavored to show throughout the book, stem from basic principles. They are stages in process. The first stage is purpose, a goal projected, and the last, the achievement of this goal.

The five intervening stages have to do with means; at first a descent into determinism to obtain formed matter to use, and then the

* The analogy can be extended. Out of the earth grow the grapes, from the grapes we obtain the wine, from the wine we distill the spirit.

employment of matter to build an organism that can achieve the goal. Almost any endeavor follows this pattern; first a goal is projected, say a business enterprise (purpose, 1), then money is raised (substance, 2), plans are drawn (form, 3), foundations laid and factory built (4), products manufactured (5), products sold (6).

But in the case of man, the house not made by hands is invisible. The only visible part of the house is the physical body. This is the formed substance—it is our fourth level. But to complete the picture, it is necessary to think of man as consisting of three other "bodies," essentially principles, at levels III, II, and I.

Hierarchy of the self

We would, therefore, chart the "anatomy of the self" as consisting of essentially four principles:

A *physical body* analogous to the fourth stage of process, essentially a collection of molecules, such being the actual substance of the body which weighs so many pounds avoirdupois.

An *organizer "body"* analogous to the fifth stage of process (the plant kingdom) in its power to create order and to organize the material. This may be the etheric body. This stage, which may be thought of as a very general kind of "*mind*" on account of its propensity for creating order, is related to the mind or ego of stage III. In the fifth substage of the seventh kingdom, it appears as "creative mind," or genius which modern man is engaged in acquiring.

An *animating "body"* corresponding to the sixth stage of process (mobility—animal kingdom). The function of this body is to provide interest, appetite, emotion, desire, and other "motivation." Like the fifth, it also has a higher form, which we describe as the sixth substage of the seventh kingdom. This higher form can be called soul; it deals in values rather than structure, and exists only potentially in man today.

The *final core of self or "spirit,"* the essential seat of consciousness, but also of evaluation, will, and conscience. At present, it too is undeveloped or potential in man, but it is of greatest importance because it is the seventh principle, whose development through several substages *is* man's evolution. There is no point in distinguishing a higher version here because whatever the seventh substage is, it is the ineffable end point of the already indefinable "spirit" kingdom.

We might also mention in this connection the preliminary results of some research into the several words, or, rather, hieroglyphs used by the ancient Egyptians to describe the "parts" or "bodies" of the self. First, and most obvious, is the *Khat*, written as a stranded fish, which stands for the *physical body*.

Then, there is the *Ka*, or double, written as a pair of arms reaching up. (According to Joan Grant, who describes these hieroglyphs in her novel, *Winged Pharaoh*, this symbol depicts *mind*, and was so written because the mind should serve the higher principles. Budge translates it as "the double.")

Best known is the *Ba* or *Ba bird*, written as a bird, but also as a human head with wings. This is translated as *soul* by both Budge and Grant.

A fourth, the *Za*, depicting spirit, is written as a circle with a mesh of lines in the form of a sieve. This is translated as spirit by both Budge and Grant.

While there are other bodies in the Egyptian hierarchy of the self, these four are the most important.

We have now described the hierarchy of selves as they stand in ideal relation to one another, an arrangement toward which they eventually evolve. But ideal government is not easily come by; it is millions of years in the making, and we will have to come down from the mountain tops to view its development a little more as it would occur in life.

Self-determination

In the preceding chapter we discussed the first five substages of the seventh kingdom of dominion. After explaining why such a kingdom must exist, we proceeded to place modern man at the fourth substage of this kingdom because of his preoccupation with combinations—combining parts to build his machines, combining persons into organizations. Complementing this preoccupation with combinations and their laws is his belief in the efficacy of such laws, his belief in science, in research, in determinism.

This brings us to the important aspect of the seventh kingdom that we have not yet discussed. It will be recalled that by a study of the seven substages we were able to isolate two properties common to them all (see Chapter VII on molecules), and hence to obtain information about the seventh kingdom. These properties were:

1. Dominion over the antecedent substages.
2. Dependence of the seventh substage upon the next higher kingdom.

The second property we have not discussed. To do so for man is awkward in this day and age. It is like discussing sex in the Victorian era. Modern man is at a kind of psychological crossroads. He is graduating from dependence upon authority and moving toward self-determination. This can be seen in a fairly long-term trend since Greek thought broke away from the sanctity of tradition and authority—typified by Egypt with its priesthood and carefully guarded ceremonies and initiations. The Greeks discovered the power of reason, and used it to assert their independence of tradition and of authority. In terms of a shorter kind, there has been in more modern times, since the American and French revolutions, a tendency to affirm the independence of the individual and to challenge "the divine right of kings." Coming down to the immediate present, we have the phenomenon of the automobile, which has given man an extraordinary freedom of movement. Add to this the freedom of the press, freedom of religion, etc., and we have an almost excessive emphasis on cutting loose from any dependence on authority.

Yet, in the long history of mankind, the dependence on authority, both enforced and voluntary, has been strikingly prevalent. Every race, every people, however primitive, have had their religions; their beliefs in supernatural beings and forces; their ceremonies for invoking the favor of the gods; their dances, rituals, temples, and idols; their myths, sacrifices, and worship.

Modern man has banished this as superstition, primitivism, the product of ignorance and savagery. And he is right in so doing. The question is whether he can, in fact, give up an age-old need without withdrawal symptoms. He has for millions of years depended on something higher, on something beyond himself, and suddenly he tries to cut this off. He cannot so easily change himself. So what results? As he has banished his worship from consciousness, it seizes control through his subconscious. He worships science. This is the most absurd distortion that man, with all his foibles, has ever indulged. For science is a tool. It provides the means for effecting his will. It should not be worshiped because its nature is service. We may not ask of science whether man has a will of his own because science is committed to the doctrine that the uncertainty which would occur if a machine had a will of its own must be eliminated. If the battery says *no* when we want it to start the car, we get a new battery. No tool would be any use if it had a will of its own. A tool is a means used by our will. So it is absurd to turn to the exponents of means, who are expert at removing self-determination from mechanisms, for any illumination of that aspect of existence which they have been at pains to eliminate.

But we cannot answer that need of man by telling him not to consult science. He is so constituted that he must have advice. He depends on something higher than himself. If it is taken away from him, he invents it; he constructs it out of the materials at hand. In an age when religion falters, he makes a cult of computers.

So we have the neurosis of modern man. He has created a brave new world of science with a thousand machines to do his bidding. Yet he has no philosophical maturity to match it. His dependence on something beyond him thus becomes more acute because he gives it no conscious outlet; he instinctively worships the computer.

The prediction is a true cul-de-sac. Science, man's new idol, disclaims responsibility for ends and goals; it will not answer questions that belong in the province of the spirit. At the same time, it manages to undermine the criteria by which man can decide moral or ethical issues. It discredits the Biblical accounts: we are not descended from Adam and Eve, but from an ape-like progenitor; the earth couldn't possibly have been formed in seven days; the world was not created on the twenty-third of October, 4004 B.C.; the Virgin Birth was a myth; Jehovah is a Jewish father image; etc., etc.

Certainly, we could hardly ask that science try to support obsolete ideas. The touchstone for science is truth. So let us have the truth, whatever the cost. But it would be most improper not to demand of these explorations into the nature of man and of process the same candor that characterizes explorations into gravitation or electromagnetism. By the same token, we would, in turn, demand the same rights that science enjoys—the right, for example, to use the techniques of mathematics, to apply geometry to meaning as well as to geography.

Meanwhile, there is always error, to which all new endeavor is liable. Let us not let the fear of making errors hold us from pushing forward. Only by making mistakes can the unforeseen be discovered.

In the preceding chapter we described what lies beyond modern man, the fifth substage, where, having learned what there is to know of law, man displays the remarkable quality we find in great creative genius: in the painters, writers, composers, leaders, and generals whose works survive their death. That this is a distinct breed different from ordinary man is not for me to prove. There is no test to establish the claim to this distinction, apart from the works of the man himself—either they live or they don't.

To show how persons of unusual and high attributes properly fit in the fifth substage, we emphasized their power of *self-multiplication* (our word that covers the two things that plants contribute—growth and reproduction). In the case of genius there is both: the figurative "growth" of their stature in the eyes of their fellows (they are "big" men), and the self-reproduction epitomized by their works, whether paintings, writings, compositions, or such historical acts as the Napoleonic Code.

We have already distinguished *between these two outlets* to creative power. Let us develop the implications. The self has now learned to draw to itself all the power it wants. And therein lies the *crise* or problem of this stage. What would happen if it drew to itself more and more and more power? Recall that it is still a finite body; it is still a person, a center unto itself, and not part of the sun. The result would be just what it would be if we tried to pass more and more electricity through a given electric light bulb. It would burn out (and, we would suppose, not just burn out a physical body, but burn out the evolving entity in a permanent sense—forever).

So here, in the fifth substage, the entity, having acquired the ability to draw all the power it wants to itself, must now find a way to get rid of this power. It must export it, *express* it in the form of creations. It is compelled, like Don Giovanni (and indeed, Mozart, himself), to be the pure creative instrument, with a list of paramours commensurate with Koechel numbers. This is the dead center of the moral issue. Before this, all other moral problems are but child's play. Here the self must really choose. And just because it can here taste the glory of power, there is the greatest difficulty for it to give up this power, or rather dedicate it, sacrifice it, if you will, to what lies beyond its immediate self-interest.

Now perhaps this seems too far-fetched or too strained a reading. Surely, we are indulging here in fantasy. Strangely, not—for this very point is one of the most completely consistent with the structure of the grid. Here, in fact, pivots our whole geometry. Recall the relation between the third and the fifth powers; how the latter goes through the same area as the former, but in reverse order. And recall that the third kingdom saw the emergence of the *identity* power, where to gain this identity the entity must create a center of its own (an enclosure for the embryo in the plant, a stomach of its own in the animal, self-consciousness in man). And recall, if you will, the story of Adam and Eve eating the fruit of the tree that was in the *midst* of the garden and therefore symbolic of the taking on of self-centeredness. Now, assuming that this process of the third power has to be gone through in *reverse order in the fifth*, as we must read the ascent through the fifth, how else can we interpret this than as expression, a *throwing off of centeredness*—which is what the plant does when it propagates itself, and what the creative genius does through his works? This is, in fact, one of the most remarkable findings of the grid, in that it gives not only a theoretical and categorical explanation of self-reproduction, but does so with great economy of means, using *in reverse* the device that accounts for atoms.

The moral question

Even more important is the fact that we have here an objective explanation of the moral question. But let us first ask: what do we mean by a moral question? What is "good"?

Clearly, the word is a very general designation for that which is suitable, desirable, satisfactory, attractive—as a good meal, a good book, etc. The *moral* issue begins, however, only when we consider ends that are beyond the immediate and local. If we go on a diet, it then becomes "good" not to eat, where ordinarily it is good to eat. The moral question begins when there is a conflict between a short- and a long-term good. "To die for a just cause" can be good if we recognize a higher self that survives such a death.

In the grid's fifth substage of dominion, we have the self with the power of unlimited growth. It has learned, like the plant, how to build order against the general trend toward disorder. It can reverse entropy! This is a tremendous accomplishment. To realize what it means, suppose we could reverse the entropy at a certain spot in space. This spot would get hotter and hotter; there would be no fuel and yet it would burn. We could, as they say, use it to light a city. More than that, it would suck in unlimited energy. This would, of course, burn out any finite vehicle. Or picture it this way. If there is a hole in a balloon, the air leaks out and the balloon collapses. But if entropy were reversed, then when there was a hole in the balloon, the air would flow *into* the balloon, and it would *not stop flowing in*. Naturally, the balloon would burst. Self-reproduction under this analogy becomes a device by which the expanding balloon releases its excess by producing other balloons.

Our reason, then, for saying that the monad must distribute its excess power is an objective, mechanical one. Infinite force cannot inhabit a finite system. We have thus correlated the moral question with something that has reality apart from morality. *We can define morality in terms of something other than itself.*

Now, a peculiar thing about morality is that in however "scientific" a manner we account for it, it haunts us nonetheless. We have shown above that it can be "defined" for fifth-stage entities. What has this to do with the rest of humanity? At first glance, nothing. We might live hundreds of lifetimes before we evolved to the stage where such an issue could come up. For it can hardly be a virtue or a credit not to do something we are not *able* to do. People who aren't able to make money try to make a virtue of this incompetence. But incompetence is not a virtue. And this applies generally. One cannot be generous if one has nothing to give. This doesn't mean that giving is an external thing, that it is the tangible gift that counts. The morality we refer to in the fifth substage is the relinquishing of the power itself, and not of the products of this power.

Putting all this together, I would reason that the moral quest achieves primary importance only for a highly evolved (fifth-stage) person. For the rest of mankind, morality is a presentiment or premonition of the importance of this at some future date—hence, the old belief in a day of judgment. The day of judgment will occur at some remote future time, when the deeds of all men will be judged. But why should there be such a long wait if there were not more deeds that we are to do in the interim? One suspects that the main point of the judgment day, even if it is a myth, is sound. There will come a time *in the lives of each of us* when we will go on or be destroyed, but this time will not come until we wield so much power that the misuse of it would destroy ourselves. But such postponement is no true escape. The self "knows" its destiny —confusedly perhaps, but with some kind of deeper insight. And it is very sensitive about piling up further indebtedness. It is scrupulous about paying for newspapers at self-service counters and telephone calls at friends' houses.

These observations should be construed as illustrations of how one can use the structure of the grid to act as a foundation for what is otherwise a slippery business—much as a sculptor uses an armature to keep soft clay from collapsing.

Sixth substage of dominion

We now come to the last substage that we can even think about or discuss. There is some question as to whether or not we should discuss it at all since it is beyond confirmation by experience, and speculation is subject to error. The reason for discussing it is that it is interesting. It is even more speculative than what we've covered so far, more fantastic. It is literally out of this world. For the entities here would not be physical. If they had physical bodies, this would be only according to some special conditions and provisions.

Another reason for discussing this substage is that it is the highest that can be touched on, and hence becomes a beacon or guiding light for the stages below it. As we've seen, dependence on the higher is of prime importance to the dominion kingdom. The sixth substage must be a realm of immortals. (This we have shown already because it is the level of energy which cannot be destroyed.) We could go on to say it is a realm of gods. Both statements plunge us into conflict with modern views. Let us treat them together.

As to immortality, the lesson that the whole sweep of the grid teaches us is that the cumulativeness which is everywhere present in the stages of the grid can be explained only by some carry-over from one stage to another that is equivalent to immortality. Further, we have seen that of the seven principles or powers, only one is "visible" (only the molecular or molar kingdom can be seen or touched or has what we call material properties). This indirectly supports the evidence for immortality, since it shows that invisibility and intangibility are no evidence for nonexistence.

As to "gods," this is perhaps a matter of definition. There are various ways we can define gods and still make our peace with the modern mind. We may define them as the elemental forces of nature, i.e., God of the Wind, God of the Sea, of the Storm, etc. In less trivial fashion we may define gods as abstract principles—as Cronus, or Saturn, is the god of limitation and law; Zeus, or Jupiter, the god of expansiveness, growth, propagation. It is only when we say that gods are actual beings that the modern mind refuses to continue the conversation. Principles, yes —persons, no!

Here is where the mind betrays us. The mind is willing to concede any abstraction, but when it draws blood, it panics. We have spread out before us a whole universe of entities all chorusing the glory of creation —and man concedes this. But he balks at the idea that there are entities higher than himself. Why should gods not be supermen—creatures which to ourselves are, say, as a horse is to an earthworm? We accept genius (or perhaps there are reservations even here—*do* we accept genius?). Why should this be the top limit? Is it because there is no tangible evidence?

Here is another turning point where the mind plays tricks with us. If we say there is evidence of higher orders of being, the mind says the evidence is insufficient, presumably because it has some reason to believe the contrary. If we ask the reason, it cites the extreme rarity of occurrence, but this means only that the ordinary occurs more frequently than the extraordinary. If frequency of occurrence is a criterion, then no prototype ever existed because there is but one prototype and there are millions of copies.

So we should not claim *tangible* evidence for the existence of immortals or beings more highly evolved than man. Our evidence is of a deductive or categorical nature, just as it was for the existence of a seventh kingdom. It is the same kind of evidence that we have for knowing that a cube has six sides, though we can never see more than three at a time.

But perhaps the modern man could accept supermen—so long as we don't call them gods. This is what I meant about blood—man is touchy on this subject.

Why is he so touchy? Is it because of what the serpent said?—"then your eyes shall be opened and ye shall be as gods, knowing good from evil." Is it because to admit this is to take on the responsibility of being a man? Is it an echo of that other masterpiece of sidestepping—the "descent from ape-like progenitors"? Why do we thrust aside the crown? We are certainly paying for it, and there is no way of avoiding the payments. The question is easy to answer. It is simply this: that the issue is so vital, so sacred, so demanding, that consciousness avoids it. While to name a thing would give us power over it, we cannot name that which has power over us!

But it is rather unfair to state the thesis in this way, and we will not enter this as an argument. It is intended, rather, as a piece of mutual soul-searching; for I am as puzzled by this as anyone.

So we have it from the grid structure that at the sixth substage of the seventh kingdom there must be immortal or god-like entities. We would suspect that there must have been many instances of these immortals living on earth—not only because of the sheer power required to launch a whole civilization or set the pattern for a whole age (Mazda, Orpheus, Christ, and Buddha), but because the accounts of early ages of all peoples state unequivocally that certain gods came down and taught the people. The Egyptian tradition has it that Osiris "abolished cannibalism, taught agriculture, built temples, invented flutes. His consort taught women to grind corn, spin flax, weave cloth."

Mayans have a god who taught writing. The Pawnees (American Indians) say Torawa sent gods to teach men the secrets of nature.

Realizing the speculative and tentative nature of this question of gods, we might enter as candidates for this category such figures from Bible tradition as Noah and Moses; but the difficulty we encounter is similar to that we meet with Osiris, Mazda, and Orpheus—we don't know enough about them.

One thing to look for would be superhuman powers, ability to perform miracles, perform healing, raise the dead, etc. Here again, not enough is known, even though the miracles of Christ are well documented. In fact, it is rather remarkable how little attention the modern mind gives to this aspect of Christ's life, considering the great number of instances of healing, raising the dead, the casting out of devils, and other superhuman power credited to Christ by the Gospels. The apostle Mark lists:

    1:25 Cures a victim of possession.
    1:34 Heals many that were sick; casts out many devils.
    1:41 Heals a leper.
    2:11 Heals a man with palsy.
    3:5   Heals a withered hand.
    4:39 Stills a tempest.
    5:8   Rids a man of many devils.
    5:23 Heals a woman of bleeding.
    5:42 Raises a child of twelve from death.
    6:40 Feeds five thousand (miracle of the loaves and the fishes).
    6:48 Walks upon the sea.
    6:56 Heals many.
    7:30 Heals the daughter of a woman.
    7:35 Heals a deaf man.
    8:8   Feeds four thousand.
    8:23 Returns sight to a blind man.
  16:6 Rises from the dead.
  16:9 Appears to Mary Magdalene.
  16:12 Appears to two disciples.
  16:14 Appears to eleven disciples.

People I ask say they do not believe in the miracles and stress their unimportance. They claim that they are fabrications and attributed to Christ because he was not an ordinary person, etc. But these rationalizations do not fit the facts. In the first place, Mark was the earliest Gospel and lists *more* miracles than later Gospels. Secondly, the miracles were the means by which Christ obtained followers and were the cause of his fame—not the other way about. There was as much skepticism then as now. An often-reported reaction to these feats was fear, which does not suggest the wishful thinking with which the modern mind clothes the era. In the case of Christ's appearances to his disciples after his death, it is clear that they required a great deal of convincing, for he had to make certain that each of the disciples had seen him (the phrase "doubting Thomas" refers to the skepticism of one of the disciples who refused to believe until he could himself insert his hands in the wounds—an opportunity he was later granted, along with a rebuke for his faithlessness).

The point in citing the miracles is not to convince the reader that they were genuine, any more than the point in describing molecules that expand and contract is to convince the reader of their existence. The point is that all of life is miraculous; and we are trying in the grid to order a number of principles, some of which we recognize as rational and therefore credible, and some of which are out of reach of the rational mind.

Seventh substage

In this substage, the dominion kingdom and, indeed, the course of evolution, reaches its goal. We cannot describe this, not only because it is so far beyond man, but because it is by definition ineffable.

XV | Process as described in myth

The preceding chapters have been devoted to constructing a theory of process. Drawing on science as much as possible for details, we have sketched in its broad outlines a theory describing the interaction of the creative with the inevitability of the laws of matter. This interaction (or what the ancients called intercourse) produces a progressive development manifesting in seven stages or kingdoms. We have not attempted to find sanction for this overall thesis from science, mainly because current science does not recognize the positive role of uncertainty in cosmology. Science, in fact, has become so fragmented into separate disciplines that it has lost sight of the unifying principle that the word "universe" implies.

Such a unifying principle was not lost on the ancients, for their speculations were motivated by a powerful urge, if not to explain, at least to describe the stages of creation and the fall of man. Their accounts in myth and legend, seemingly naïve, have an amazing sense of wholeness, of integrity, and contribute in a way that science, with its emphasis on the explicable and on the detailed development of successful techniques, has lost. Science, like a map, can furnish information, but it cannot provide a compass. Myth supplies this compass. With its help we can discover how to orient the map.

There is no compulsion in this—everyone can still go where he pleases. We should realize that the compass reading that orients the map does not dictate or even concern one's own personal destination; it has only to do with how to orient *the map*, which is essential for any use of it. Myth can help with the orientation we seek, for it is in rapport with nature with which modern man has lost touch. This rapport with nature—with the unconscious; with the mysteries of life and death, of generation and transformation; with that area of knowing which linked man with life instead of holding him off, separately, as an observer— permeates the literature handed down to us in myth and legend, in art and symbol.

Approaching myth

Many important myths deal with the descent and ascent of man. If we include cosmologies as a pertinent and necessary preface for the descent, and hero myths as dealing with the ascent, unquestionably the more important myths do fit the description. Indeed, if we get behind their superficial expression, their content is surprisingly profound and rich in meaning. Perhaps that is why they have survived. They echo in palatable form the deeper truths of existence and have survived because of this alignment with universal meaning. At this point a difficulty arises. Who is to interpret this meaning? For myths, of course, must be interpreted—as, indeed, they are, and in quite different ways.

Old interpretations

Here it is interesting to note the way in which Plutarch, as G. R. S. Mead* pointed out in 1906, considered the various theories of his day which professed to explain the ancient myths and theologies. Among them was Euhemerus' theory that the gods were nothing but ancient kings and worthies. Plutarch dismisses this as an insufficiently satisfactory explanation. The theory that gods referred to "daimons" (as exemplified in Homer when the gods inspire men to act in certain ways, or otherwise assist, punish, or reward them) he considers to be an improvement. He also takes into account the theories of the physicists or natural phenomenologists (who claim that Osiris represents the Nile, etc.) and the "mathematicians" (who think of the gods as references to the heavenly bodies, Osiris to the moon, etc.). From all of these, Plutarch

* Mead, G. R. S. *Thrice Great Hermes* (vol. I, pp. 257, 318). London: John M. Watkins, 1906.

concludes that no simple explanation by itself gives the right meaning.

"But," to quote Mead, "of all the attempted interpretations, he [Plutarch] finds the least satisfactory . . . those . . . content to limit hermeneutics (explanation) of the mystery myths simply to the operations of ploughing and sowing." With this "vegetation god" theory Plutarch has little patience, and stigmatizes its professors as that "dull crowd."*

The new: Freud and Jung

Now, seventy years later, we have some new schools of interpretation, the psychological, both the Freudian and the Jungian. Freud's attitude amounts to a dismissal of myth from serious consideration. It reduces all symbolism to the level of sex (instead of raising sex to a universal principle of creativity). It is true that Freud made an important and necessary contribution by exposing the hypocrisy of Victorian attitudes. But in regard to a myth which openly states that the phallus of Osiris is its central theme, no such crusader is necessary. In fact, the myths about Osiris—and, for that matter, the Greek myth of Uranus—are so patently sexual that we may reasonably suspect the sex symbols conceal a deeper meaning, and thus reverse the situation to which Freud's theory of censorship applied.

Coming to the Jungian interpretation, we draw closer to the true content of myth. Jung, building on the rubble left by Freud, recognizes a deeper layer, the collective unconscious, which is a repository not only for the past or repressed memories of the individual, which lurk in Freud's subconscious, but for the much more universal *race* memory. Here, according to Jung, lie all the memories of mankind in a kind of universal sleep, raising themselves from time to time in dreams or flashes of surmise, possibly to warn us at some important crisis, but in any case available as subliminal undertones that enrich us and provide that curious resonance from the inner being that makes it possible to "recognize" the true and the good, as well as to be invigorated by old tales and legends. With regard to sex, the Jungian theories again go a

---

* Plutarch's criticism applies to many of the current theories stemming from Frazer, who in *The Golden Bough* promotes the "vegetable god" theory.

step beyond Freud by elevating the feminine symbol to the status of the soul rather than the immediate object of animal passion.

We cannot, however, rest here. The Jungian concepts, though they go a long and important way toward it, do not reach the true center of the myth. A more extended study of the language of symbols reveals a much greater variety of meaning than it is accorded by most Jungians. In the light of the theory of process, the archetypes comprise a whole dimension (level) of existence. They are the inhabitants of the psychic world and the variety of their manifestations is almost unlimited. This variety runs the gamut from universal archetypes to personal dream symbols, cartoons—even ordinary language. They provide the substance of life.

The translation of symbols tends to be limited by the range of understanding of the translator. We, of course, run into the same limitation; but progress is often like the group velocity of waves, in which the individual waves arise at the rear of the group and push forward until, just as they are merging at the front, they fade and disappear, contributing to the general motion but vanishing as they get "too far out."

The grid theory and myth

Our thesis at this point is the pertinence of myth to grid theory and of grid theory to myth. Readers still skeptical of the validity of grid theory will naturally enter this new territory with a certain reserve, withholding judgment until further proofs are furnished.

Unfortunately, this reasonable reticence will not make any sense in the areas we have entered, where proof is especially hard to come by. This chapter is not intended as a proof of grid theory. Such proof as exists of grid theory has already been given in earlier chapters. Since we are therefore assuming that grid theory is more or less valid, the function of this chapter is to draw on myth for guidance in the application of grid theory to man, for orienting the map. We propose, therefore, to go ahead and assume a correspondence between grid theory and myth. Our justification is that we are thus able to draw from myth greater meaning and pertinence.

There is, however, besides the grid theory, a second hypothesis whose validity must be assumed: that primitive myth was in rapport with the basic workings of the cosmos and did correctly depict cosmological truths. This hypothesis does not state why myth should be correct. It does not tell us whether man invented fantasies which had validity because his intuition perceived the truth, or whether there might not have been in some remote age great leaders who taught theology and cosmology in a form that could be grasped and retold by simple people.

For example, even the quite remarkable instance of Swift's description of the moons of Mars is susceptible to a double interpretation. Swift, in *Gulliver's Travels*, explains that Mars has two moons, and gives their period of revolution as very short—quite close to the actual periods of 7 hours 39 minutes, and 30 hours 18 minutes— which is not what one would expect on the basis of the fact that our own moon takes 27 days to revolve. Yet Swift wrote his fable two hundred years before the moons of Mars were actually discovered by Asaph Hall in 1877. Did Swift "intuit" this remarkable bit of information, or did he learn it from some ancient or forgotten source? Some maintain the latter, instancing the fact that the Mars of mythology has his chariot drawn by two steeds, Phoebus and Deimos (the names, incidentally, which astronomers later chose for the red planet's two moons).

In this connection, I will mention here that I have seen references to the effect that the ancient Hindus held that neither Mercury nor Venus rotates on its axis with respect to the sun. If this is true, it implies a knowledge of astronomy surpassing that of the present time, for modern astronomy, with the help of present-day instruments, has only recently confirmed this lack of rotation for Mercury. Furthermore, temperature measurements of the clouds of Venus disclose that its period of rotation is very slow and thus confirm the Hindu teaching. Thus ours may not be the first advanced civilization and, if this is so, myth may be the remnant of ancient teachings rather than the result of intuitive wisdom.

Apologia for symbols

As we have said, myths demand interpretation, and our hypothesis is that, since the universe is governed by and exhibits the attributes of process, myth will recount this process in symbolic form. The question might still be raised: why symbolically? why not directly? This is an interesting question. It thrusts back into the nature of language itself, for what would a direct description of cosmology be? Recognize, if you will, the great difficulty of the physicist in describing the atom. First he had the picture of a billiard ball; then of a minute solar system, with electrons like planets flying around a central sun; then the Bohr atom, with orbits mysteriously regulated by quantum laws, revealing that they must be in some special numerical relation to one another; then the orbits giving way to a "probability fog." The effort of the physicist to correctly interpret the atom, we noted in Chapter VI, is like the effort of Scripture to describe the truth of revealed religion: both have to draw on the sensible world for their images.

This is also true of a cosmogony, whether the beginning be described as the creation of matter out of light, or as the intercourse of Sky and Earth, or as the separation of the waters above from the waters below. The images and words used are only a device to assist the mind. There are no actual "things" at the level of electrons or in the first steps of creation, and there is no image except one that will have to be translated or erased. This is where the language of symbolism triumphs in the end. It reminds one of that wonderful story, now decades old, about Noel Coward sending a telegram and signing it "Mussolini." When the telegraph operator saw the signature, she cited a ruling that false signatures were not allowed. So Noel Coward signed it "Noel Coward." The operator rebuked him again. He replied, "But I *am* Noel Coward!" "Oh, in that case," said the operator, "it will be all right for you to sign it 'Mussolini.' "

Herein is the defense of anthropomorphism. We translate the meanings we discover in nature into more and more abstract entities and then, at last, realize that however we translate them, we can understand only that which is human, and we might just as well say the electron "attracts" the proton. But let us examine symbolism further.

A symbol is something that stands for something else and, according to the modern view, the assignment is arbitrary. Thus, while in algebra the letter x stands for the unknown, any other letter would serve as well. In ancient times, to the contrary, there was an immediate and nonarbitrary connection between the symbol and its meaning, thus rendering translation possible. Indeed, the language of symbolism, like the language of dreams, can be translated precisely because the assignment of symbols is not arbitrary, but guided by the inherent rapport between the abstraction and the object symbolizing it.

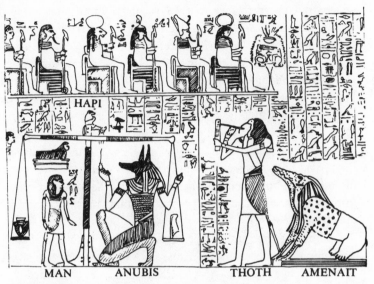

MAN   ANUBIS   THOTH   AMENAIT

The weighing of the heart or soul of the dead (Egyptian)

Let us give as an example the painting representing the judgment after death. It depicts the weighing of the heart or soul of the dead. In the picture, the dog-headed god, Anubis, operates the scales; the mandrill god, Hapi, makes the pointer reading; and Thoth, the ibis-headed god, records the result. The "heart" is weighed against a white feather, and Amenait, a hybrid monster with the head of a crocodile and the rear of a hippopotamus, crouches nearby waiting to devour any leftovers disqualified by the test. The message is quite clear even without translation. But it is interesting to know some of the particulars. Thoth, whose function is depicted not only by his holding a pen and writing down the results, but by the ibis beak springing from his head, symbolizes the power of the mind. Curiously enough, the name Thoth is close to our word "thought," which describes his role. But the Egyptians, having no abstract words like "thought," gave Thoth an ibis head to symbolize his function. Anubis, the dog-headed god, is man's faithful helper and guide. With his exaggerated nose and ears, depicting heightened sensitiveness, Anubis represents the powers of discrimination. Hapi, who watches the pointer, has the head of a mandrill. To anyone who has observed a mandrill and noticed the extraordinary power in its concentrated gaze, this way of emphasizing the accurate reading of the pointer is brilliant.* The white feather symbolizes truth, for the most interesting reasons. As we shall later see, air is, in general, the symbol of mind (as we use the term "to air" in the sense of making known), and as a feather moves air, so does truth move mind. The crocodile-headed Amenait represents the physical world that recycles the debris.

Contrasting with this very ancient "household of the self," we might instance Walt Kelly's Pogo. Pogo's immediate circle includes Albert, an alligator, Owl, and Churchy (from *Cherchez la femme*), a turtle. Now Pogo, ostensibly a possum, is drawn with a round face from which lines radiate, a symbol that everywhere represents the solar or central principle, the spirit itself. Owl is clearly mind, and Albert, like Amenait, the physical (body).

Comparing Pogo and his friends with the four functions of Jung —intuition, intellect, sensation, and emotion—we note that all are accounted for except emotion, so Churchy must be correlated with this function. This is borne out by Churchy's fondness for song and from his name, Cherchez la femme. But why should emotion be represented by a turtle? What is a turtle? A turtle is an animal with an armored shell into which it withdraws under threat of danger. So the story depicts modern man as wearing an emotional armor. Again, Pogo, or spirit, is depicted as a possum since a possum is an animal that pretends to be dead.

Now, I'm sure Kelly had no such roles in mind when he created Pogo and his friends. Indeed, if he had had such a notion consciously, it would probably not have emerged with the telling conviction that the unpremeditated version has. Whether in Kelly's cartoons or Egyptian papyri, we have before us a language that is in rapport with the basic truth of nature, not because it is highly conscious, but because it is not. It is as unconscious as digestion or those other physical processes that

---

* In some versions, the mandrill has degenerated into what seems more like a decoration on top of the scales than an active member of the team.

our body is able to carry out with no assistance from the mind—such as the healing of wounds, immunization against disease, growth of an embryo and, for that matter, all growth. This instinctive functioning is perfect and provides a sort of built-in compass that guides the "natural" person, just as the sense of equilibrium, established by a mechanism in the inner ear, enables us to walk upright.

The grid and myth

Our theme is that myth, in general, symbolically describes the arc of process, which begins with the "descent into matter." In the grid, process begins with undefined purpose or impulse, initially completely free, and then takes on a series of limitations that eventually tie it down to complete determinism. In myth, the synthesizing or return half of the arc is sometimes a continuation of one and the same saga; in other cases, it may be described in separate myths.

The real difficulty is in translation. Not only are myths in a special language, i.e., the language of symbols, but in the final analysis, *there is no language for the ultimate nature of things*. The physicist resorts to mathematical formulae, but never really knows what he is talking about. He trains himself to avoid visualization, to navigate through the darkness by the use of instruments. Only occasionally does some flash of genius —like a stroke of lightning—illumine the path for a moment and permit a new sighting to be made on the goal.

Translation, then, is our problem. Among other things, we must also translate what we mean when we say the universe is process. Perhaps the most apt expression of this thesis is the formula of quantum theory due to Dirac (a and b are operations, h is Planck's constant, i is imaginary):

$$ab - ba = ih$$

This is the celebrated equation which expresses the breakdown of the law of commutation; that the operation a times b is the same as b times a, or $ab = ba$. But even this has to be translated; it is not enough to say that $ab - ba$ doesn't commute. We must so translate the formula that every element is accounted for. Clearly, ab is one operation, and ba is its inverse, like going to town and returning. Going to town and returning gets us back where we started, so the result is zero geographically. But when in town, we signed a contract, we *did* something, and this was in no sense a geographical change of position. That it is not geographical is signified by the letter *i*, which in the quantum formula represents the square root of -1. It describes the action of *h* as of a different nature from the change of position described by *ab* and *ba*. The *i* is imaginary, that is to say, in a different dimension, just as signing a contract is different from moving from place to place, since one is a value transaction, the other a physical transaction. So our reduced formula, $ab - ba = ih$, says that process is an involvement with matter (*ab*) and an evolvement from matter ($-ba$), to produce a "nonmaterial" *unit* of action. The matter, then, is means; the imaginary part, the end achieved, as in the arc.

The beginning of things

Judaic

It is our thesis that myth, in its complete form, says the same. To begin on familiar ground, let us start with Adam and Eve. Adam is the first man, *the first cause*. By itself, first cause is as nothing; it lacks something. So to it is added Eve, "the mother of all living," the *desire principle* (Eve, as desirable, epitomizes desire). Then comes the step that makes them conscious of themselves and thus responsible for their acts—the eating of the fruit of the *tree of knowledge* of good and evil, the deed that will make them wise as gods, and represents their involvement, their contract with necessity, their enrollment in the school of life. Next follows their expulsion from the garden, their descent into the *world* where, the Lord says, "in sorrow shalt thou eat of it (the tree) all the days of thy life." But we still do not have the whole story. The emergence, or evolvement from matter, is not described in Genesis. Perhaps it is best exemplified in the New Testament by Christ himself, the hero who suffers crucifixion in matter and emerges triumphant. And it is also exemplified by Christ in his teaching of *rebirth*. Rebirth in Christ is, again, a symbol. It describes awakening, the consciousness of the divine

within the self and of our ultimate potential, the capacity to be sons of
God.

Egyptian

Such interpretation brings another curious and important myth into
place; the Egyptian myth of Osiris, darling of the gods, against whom
his brother Set plots downfall. Set prepares a magnificent casket,
encrusted with gold and jewels, to the exact measure of Osiris. Then
he invites all the gods to a party, promising the casket to the one who
fits it best. Each of the gods tries it, and when Osiris gets in, Set slams
down the lid and throws the casket into the Nile. The waters of the river
carry it as far as Byblos, where it comes to rest by a tamarisk tree, which
eventually grows around the casket, enclosing it.

This double emphasis on enclosure and restraint (first by the casket,
then by the tree) may be correlated with the progressive loss of freedom
we have come to expect at the second and third stages of process. The
next development, in which Osiris' sister, Isis, obtains possession of the
casket, only to have Set discover it and cut up Osiris' body into fourteen
pieces, which he scatters in the marshes, emphasizes that now freedom
is completely lost: Osiris is disintegrated and descends into *disordered*
fragments. This correlates with the fourth, or deterministic, stage of
process, where the entity is fractured into parts lacking self-energy.*

The next part of the myth begins the "return." Isis searches for the
fragments of Osiris and finds all but one, the organ of generation. She
nevertheless joins the fragments together, reanimates the corpse and,
by union with it, conceives a son, Horus. Horus, represented with a
falcon's head, returns us to the "higher" or upper level, from which
Osiris descended. Horus is also the infant Sun God, reborn every
morning and, as a manifestation of Ra (the Sun God), returns the cycle
to where it commenced (since Osiris was the son of Ra).

The reason for the overlapping manifestations of Horus lies in the

---

* To define high and low in terms of unity versus fragmentation is probably more
basic and valid than in terms of literal height, because order–disorder is a true
"invariant" and not dependent on the arbitrary direction of gravity.

fact that he is both the beginning and the end of the cycle whose middle is, in effect, Osiris, God of the Lower World, the dismembered man–god.

What is most interesting here is that the detail of the lost member of Osiris, and the conception of Horus, is another way of describing a virgin birth. Neither Christ nor Horus has a physical father, a prior cause, and the great lesson these myths teach is that the final essence, the ultimate cause of life, is not a thing, but is cause itself. When we come to the complete fragmentation and entombment in matter, to the rock bottom, we cannot depend on any outside thing to lift us up. *We* must do it ourselves and, in that act, we are reborn.

Greek

Our next example is that most wonderful of myths, the Greek account of the beginning of things. Actually, this is more a cosmogony than a myth of man, but since both man and the universe are processes, the Greek myth may be profitably examined in juxtaposition to the two we have just considered. It tells that in the beginning, Gaia, or Mother Earth, was made pregnant by Uranus, God of the Sky. So burdened was she with frequent childbearing that she craved relief and, to this end, gave her son, Cronus, a sharp sickle and persuaded him to use it. Cronus did so and cut off the testicles of Uranus and threw them into the sea. From the blood came the Furies, and from the sea foam was born Aphrodite, or Venus. Cronus became king but, having been told he would in turn be overthrown by a son, he ate his own children as fast as they were born. His wife, Rhea, by the stratagem of wrapping a stone in swaddling clothes and presenting it to Cronus, succeeded in saving her son, Zeus, who overthrew and succeeded his father as predicted.

In Cronus, we have the principle of determinism, which cuts off and *regulates* not only his father, but his own children. We would, therefore, relate Cronus to the fourth principle, and Zeus to the fifth, since he escapes from the regulation of Cronus as vegetation "escapes" from determinism through its progeny. Note the emphasis on stratagem as the way to get around the "law" of Cronus. "Stratagem" is another way of viewing this "escape" to regained freedom. The "Wily Ulysses" is the Greek representation of the hero who escapes the restraint of determinism. This may seem a far cry from the Christian virtues until we recall that Christ as a child was spirited out of Bethlehem to escape the decree of Herod.

The seven stages

Greek

The Uranus–Cronus–Zeus myth, which is an account of the generation of the universe, lends itself to interpretation in terms of the actual stages of process. The myth depicts a succession: Cronus is the son of Uranus, and Zeus is the son of Cronus. Investigating further, we find that Uranus is the son of Gaia (Earth). Note the sequence:

				Stage
Gaia	Mother Principle	=	Substance	2
Uranus	Son of Gaia	=	Seed Principle (identity)	3
Cronus	Son of Uranus	=	Determinism	4
Zeus	Son of Cronus	=	Escape from determinism	5

Gaia, or Mother Earth, conforms to substance. She supplies the substance, and correlates to Eve, "the mother of all living," in the Adam and Eve story (Genesis 3:20).

Uranus' seed impregnates Mother Earth. This seed principle correlates with identity, stage three.

Cronus eats his own children, and thus he represents limitation by law, or determinism. In emasculating his father, Cronus had similarly delimited his father. Cronus is time* (chronometer, chronic, chronology), or, again, Father Time with his scythe.

Zeus' escape from being eaten by Cronus, from limitation by time, correlates with the fifth stage of process, the power of vegetation to project itself through its seed, and thus conquer time. Zeus is also known for his progeny.

Zeus, however, is but one of many heroes in Greek mythology. Hercules, Jason, Theseus, Perseus. Each one is significant in the context of the arc.

---

* Francis Bacon gives this interpretation in *Wisdom of the Ancients*.

The Greek myth of creation gives us no clear correlation to the first stage, though we can definitely say that there was something before Gaia. Some accounts give Chaos as the origin of all. To quote Kerenyi:

> Ancient night conceived of the Wind and laid her silver egg in the gigantic lap of Darkness. From the Egg sprang the sun of rushing wind, a god with golden wings. He is called Eros, the God of Love. But this is only *one* name, the lowliest of all the names this god bore.*

It would be consistent with other accounts if we could claim that Eros set the whole thing in motion and was the father of Gaia, but we cannot. Hesiod places Chaos first, with Eros born after Gaia, as a brother or co-equal with Gaia. Perhaps the Greeks, as the first materialists, just did not give to light and fire that primal function that is their right.

Judaic

To obtain a correlation with the first stage, we may turn to the account in Genesis:

1:1 In the beginning God created the heaven and the earth.
1:2 And the earth was without form, and void; and darkness was upon the face of the deep. And the Spirit of God moved upon the face of the waters.
1:3 And God said, Let there be light: and there was light. . . . the first day.

Iranian

The Iranian legend of the beginning of the world is as follows:

> In former times the two realms of Light and Darkness . . . constituted a complete and balanced Duality. This equilibrium was disturbed when the Prince of Darkness was attracted by the splendor of the realm of Light. Before the threat of his onrush the Father of Greatness evoked a number of hypostatic powers of light. Their defeat and subsequent disappearance into darkness lies at the origin of the state of mixture.**

* Kerenyi, Carl. *The Gods of the Greeks.* Translated by Christopher Holme. London: Thames & Hudson, 1951; New York: Grove Press, 1960, and Greenwood Press, 1962.
** Dresden, N.J. "Mythologies of Ancient Iran." In *Mythologies of the Ancient World.* Edited by Samuel Noah Kramer. Garden City, N.Y.: Doubleday, 1961.

The word "hypostatic" means "having substance," so the ancient myth seems to be describing an aspect of light which has only recently come to be recognized, i.e., that light exists in the form of finite bundles or photons, each having a particular energy.

Dresden goes on to say that "among the first hypostases which were absorbed by darkness was primeval man or Ohrmizd, as he is known in Iranian sources."

The complete Iranian account as Dresden gives it* can be correlated to the grid stages as follows:

	Stage
First, he created the sky, bright and manifest . . . in the form of an egg of shining metal. . . . The top of it reached to the Endless Light; and all creation was created within the sky. [The expression "reached to endless light" seems to anticipate modern ideas of the electromagnetic spectrum.]	1 (Light)
Second, from the substance of the sky he created water. . . . [We have frequently mentioned the resemblance of nuclear particles to water and substance.]	2 (Substance)
Third, from the water he created earth, round, poised in the *middle of the sky*.	3 (Having a center)
(Number omitted) And he created minerals within the earth. . . .	4
Fourth, he created plants. . . .	5
Fifth, he fashioned the . . . bull. . . .	6
Sixth, he fashioned Gayomart (the first man). . . . (The seventh was Ohrmizd himself.)	7

In this Iranian account, plants actually are numbered as coming fourth and animals fifth, instead of at the fifth and sixth stages. Minerals fall in the fourth stage, but are not numbered. It would be my opinion that the writer confused two of the earlier stages and combined them because, although the names are different, the descriptions resemble the corresponding stages of the grid. The third stage, although called

* *Ibid.*, pp. 338–339.

"earth," is described in a third-stage way ("round, poised in the middle of the sky . . ."), which sounds like the third stage of the grid—"having its own center." Since the next numbered stage is the creation of plants, the stage that is missing is that of minerals, which we include in the molecular kingdom.

Mayan

Less widely known, but one of the most interesting, is the myth in *Popul Vuh*, the 16th-century manuscript, written in the Quiché Indian language, which records fragments of the mythology of the Mayans.* As the *Popul Vuh* recounts it, the story begins with the twin brothers, Hunhun-ahpu and Vukub-Hunhun-ahpu playing ball in Heaven. The twelve Princes of Xibalba (gods) send their four owl messengers to Hunhun-ahpu and Vukub-Hunhun-ahpu, ordering them to appear for their initiations. Failing these, the two brothers pay with their lives, and the head of Hunhun-ahpu is placed in the branches of the sacred calabash tree, which becomes laden with luscious fruit. Xiquic, the virgin daughter of Prince Cuchumaquic, learns of the sacred tree and, desiring some of its fruit, journeys to it. When Xiquic puts forth her hand to pluck the fruit, some saliva from the mouth of Hunhun-ahpu falls into her palm and the head speaks to her, saying, "This is my posterity. Now I will die."

The young girl returns home. She becomes pregnant and is questioned by her father, who refuses to believe her story. At the instigation of Xibalba (the gods), the father demands her heart in an urn. Xiquic persuades her executioners to spare her life. In the urn, instead of her heart, they place the fruit of a certain tree whose sap is red and has the consistency of blood.** In due time, she gives birth to twin sons, Hunahpu and Xbalanque, who grow up and do great deeds. The Princes of Xibalba hear of them and summon them to the mystery initiations, which take seven days and are intended to destroy them. These are the same initiations at which their father failed, but Hunahpu passes all the

* *Popul Vuh*. Translated by Delia Goetz and S. G. Morley. Norman: University of Oklahoma Press, 1950; London: Wm. Hodge & Co., 1952.
** Note here too the ruse to escape the law of the fourth stage.

tests and Xbalanque fails only in the last, in the Cave of Bats, where his head is cut off by the King of the Bats. Hunahpu, however, has by this time attained magical powers and restores his brother to life.

Having passed their initiations, Hunahpu and his brother become itinerant magicians! They go about giving performances in which they do extraordinary tricks: one carves the other into pieces and puts him back together, and so on.

This is most interesting, for it supplies what is often neglected in other myths, the sixth stage. Of the fifth we have had many examples: it is the birth of the hero and his "passing of tests" (as with Hunahpu and Xbalanque), the twelve labors (as with Hercules), the slaying of dragons (as with Perseus), etc. The seventh, or final, stage is where the hero reaches god-like status (Horus becomes the Sun God, Christ "sitteth at the right hand of God the Father," and Hunahpu and Xbalanque become the Sun and the Moon). But the sixth is not often clearly enunciated.

Let us recall what character the sixth stage, by its relation to the second, must have. The second is attraction, the *spell* of illusion.* So we could expect the sixth to be similar; but in the case of the *Popul Vuh*, the self *projects* illusion, rather than being entrapped by it. We have referred earlier to "transformation" as the more generalized reading for the mobility appropriate to this stage and in the reference in fairy tales to the magicians who change into mustard seeds, and into hens who eat them, etc. In any case, in this myth we have magic or the *creation* of illusion, the inverse of entrapment by illusion.

To return to Hunahpu and his brother, their performances of magic reach the attention of the twelve Princes of Xibalba, who invite them to perform. After causing the palace of the princes to vanish and reappear, they cut up the pet dog of the princes and restore it to life again. Intrigued, the princes ask if they could be cut up and restored. The brothers assent, and cut up the princes, but do not restore them!

This concludes the drama. Hunahpu and Xbalanque become the celestial bodies, the Sun and Moon.

* Mircea Eliade, in *Images and Symbols* (translated by Phillip Mairet, New York: Sheed & Ward, 1969), gives a full chapter to "The God Who Binds." Binding for Eliade is by magic spells, but we generalize the concept to include the loss of freedom due to "attraction by substance," that is, the second stage.

Process as described in myth

Light	Substance	Identity	Separation and combination
Genesis, chapter 1 ". . . and God said, Let there be light."	"Let there be a firmament . . . . Let it divide the waters from the waters."	"Let there be gathered to- gether in one place . . . every- thing whose seed is in itself."	". . . lights to divide the day from the night . . . seasons . . . days . . . years . . . to rule the day."
Old and New Testaments Adam.	Eve, "mother of all living."	The tree of knowledge of good and evil . . . in the midst of the garden.	Toil and sorrow, banishment from the garden. Virgin birth of Christ.
Egyptian Osiris.	The jewel-en- crusted casket floats down the Nile.	The tamarisk tree encloses the casket.	Dismemberment of Osiris by Set . . . Isis gathers the pieces. Vir- gin birth of Horus.
Greek Eros?	Gaia (Mother Earth).	Uranus, the Sky God son of Gaia.	Cronus, son of Uranus and Gaia . . . eats his own children (time). Birth of Zeus.
Iranian ". . . first the sky . . . the top of it reached to endless light."	". . . second, from the sub- stance he fash- ioned water."	". . . third . . . earth round, poised in the middle of the sky."	Fourth stage omitted?
Popul Vuh Twin brothers.	Forced to take initiations by the twelve Princes of Xibalba. They fail.	Hunhun-ahpu's head placed in calabash tree.	Sacrifice of Xiquic de- manded . . . she escapes by a ruse.

Growth	Transformation	Dominion
"Let the waters bring forth abundantly . . . be fruitful and multiply."	". . . cattle, and creeping thing, and beast of the earth (man–dominion)."	". . . and God resteth the seventh day."
Christ's resistance to temptation by Satan.	Miracles. Resurrection . . . appearance to disciples.	"Sitteth at the right hand of God."
Horus grows up and becomes the hero who conquers Set.	Trial for possession of the eye.	Horus, the Sun God.
Zeus escapes edict of Cronus. Many heroes.	Children of Zeus, Ares, Minerva, etc.	
". . . he created plants."	". . . he fashioned the bull."	Gayomart.
Virgin birth of Hunahpu and Xbalanque. They pass tests.	Brothers become itinerant magicians. They fool the twelve princes.	Brothers become the Sun and Moon.

	Stage
The twin brothers play ball in Heaven.	1
They fail their initiations (deceived by illusion*).	2
Hunhun-ahpu's head is placed in the calabash tree and the Lords of Xibalba say, "Let none come to pick of its fruit."	3
The maiden, Xiquic, comes to pluck the fruit and becomes pregnant. By a ruse she escapes execution.	4
Twin sons are born to her, Xbalanque and Hunahpu. They take the initiations and succeed.	5
The twins perform magical tricks and deceive the twelve gods, cutting them up and not restoring them.	6
The twins become the Sun and Moon.	7

Summary

Comparing these several accounts, the resemblance is especially striking in the recurrence of the tree at the third stage. This is curious because the tree in each case seems to have a different meaning. In the myth of Osiris, the tree grows up around the coffin. In Genesis, it is the tree of the knowledge of good and evil. In *Popul Vuh*, it is the sacred calabash tree in which the head of Hunhun-ahpu is placed. The only explanation I can find for this reference to a tree is to draw on the theory that at this stage, process becomes capable of relationship (this is the form, or conceptual, stage), and relationship is expressed by a tree, as in the term "family tree." This is the third stage, spirit trapped in mind, represented by a tree because a tree with its many branches (ramifications) suggests the many kinds of relationships with which the mind deals.

The "trap of mind" is exemplified by the myth of Perseus slaying the Medusa. Medusa is depicted with a head from which snakes grow like hair (the powers of mind). Her effect on people is to turn them to stone (the mind "objectifies," i.e., makes inert). To deal with this difficulty and avoid being himself turned to stone, Perseus looks at her in a

---

* In the first test the brothers are deceived by a wooden figure in the likeness of one of the gods. Note that in stage six, this is reversed; the twins deceive the twelve gods.

mirror, itself a symbol of mind ("Mind is the slayer of the real; only the mind can slay the slayer," as the teachings of Zen put it).

At the fourth stage (complete loss of freedom) there is emphasis on difficulties ("in sorrow shalt thou eat of it all the days of thy life," Genesis 3:17; in the Greek myth, Cronus eats his own children; in *Popul Vuh*, the princess Xiquic must be sacrificed). The problem is resolved only by the virgin birth of the hero. In the Osiris myth, Isis conceives Horus from the corpse of Osiris; in *Popul Vuh*, Xiquic is impregnated by a dead head. But this similarity is not imitative, is not such as to suggest a transmission from Egypt to America. Superficially, there is no resemblance between the myths. It is only at the deeper level, when correlated to process, that the similarity emerges.

To discover this basic correspondence, let us note that all these examples, including the cosmogonies, deal with a *descent*, followed, after a virgin birth or a ruse, by an *ascent* or, in the case of Cronus, an escape from limitation (determinism). Thus the birth of Zeus, famous for his amours and his progeny, restores the power of generation that Cronus terminated in Uranus.

I	1 Freedom (potential)	7 Freedom (actual)
II	2 Binding	6 Unbinding = Motion
III	3 Form	5 Growth = Zeus
IV	4 Determinism	

Again in the Osiris myth, Horus at stage four conquers Set, the principle that first trapped Osiris and imprisoned him in the jewel-encrusted casket at stage two. In *Popul Vuh*, the twin sons pass the initiations their father failed. As magicians, they cause the Lords of Xibalba to become victims of the same weapon (illusion, stage six) that was used against their father at stage two.

Thus we can place the seven stages of myth on four levels and discover that at each level the right-hand side frees itself from the limitation that arose on the left at this level. The levels for myth too have the meaning we found for the kingdoms: level I is freedom, level II is binding, level III is form, and level IV is determinism.

XVI | The evolution of the self

Man's evolution

Quite apart from the validity of our theory of process, surely the exercise has highlighted an important subject. We become aware that present theories of evolution have nothing whatever to do with the evolution of man, by which I especially mean the evolution of *individual persons*, which is the issue of greatest importance for each one of us.

In chapter XII we dealt with the inadequacies of familiar evolutionary theories at some length, and it should be obvious that the "evolution of species" cannot apply to man, for all races and colors of man are of one species. One of the aims of Darwin's theory was to account for the many different species of birds he encountered on the Galapagos Islands of the Pacific. He speculated that all these different species had descended from a common ancestor. But the evolution in this case operated to create different species from one more primitive type. A species is defined as a group distinct and different to such an extent that members of different species cannot interbreed and produce fertile offspring. There are six other increasingly broad classifications used by biologists:

Species	Lion (*leo*)
Genus	*Felis* (includes leopards, etc.)
Family	Cat (including lynxes)
Order	Carnivore (includes bears, dogs, etc.)
Class	Mammals
Phylum	Vertebrates
Kingdom	Animal

But all mankind is only one *species*, the narrowest category of all.

Moreover, as we noted in Chapter XII, there has been no significant change in man's physical body over recorded history (eight thousand years). While this has been a short time as evolution goes, the evidence in this period has not indicated that survival of the fittest has had any trace of significance so far as changing man's physical body or genetic structure. True, disease and famine may have eliminated the less healthy, but this has not changed man's physique.

Civilization has interposed itself between man and the "laws" of survival by establishing protection partly by separation of function—soldiers, workers, etc.—even to the extent of breeding from the weak; and survival of the tribe replaces survival of the individual. A civilization doesn't succumb because members of its population are unfit, but because the civilization becomes decadent, a phenomenon not to be accounted for in terms of gene structure.

But all these considerations have no bearing on what is really important, your evolution and mine. Here science merges into religion, for we all go about the business of living with a complete seriousness as though we believed that our every act contributed to our total development. It seems to me that we know in our bones that gradual and continued growth of character and competence is more important to us than any other end and that our belief in Christian doctrine is sustained by our participation in evolution. Faith in a heaven and in an ultimate reward for a good life does not owe itself to the Christian Fathers or to the Old or New Testament, but to the fact that evolution has pointed toward a better future for billions of years.

Of course, I am writing about a theory of process, which is to say, of evolution, whose thesis is that we have evolved through billions of years, perhaps even through many universes, from photons and atoms through molecules, cells, and ultimately through animals to reach the stage at which we could be born human and start learning to talk at two years of age and in some cases write symphonies at seven.

While such a theory is straightforward and conservative in the sense that it requires a minimum of assumptions, the reader may find it unconvincing and prefer to await its support by the authority of science.

But I question that source for conviction. Conviction should come from within, and it is my hope to awaken it in the reader. The evidence for it is the concern we have for self-improvement and for long-term goals; the utter importance to us that we be right; the great effort we make, when we are convinced we are not right, to reform; and perhaps most important, our belief in what is better than we are, what is beyond our grasp.

And it is this orientation that has provided the force to bring us to our present stage, which is a long way, a very long way, from where evolution started. For we are not just a bunch of atoms; we are generals of an army or, if you prefer, leaders of an organization whose membership is a billion billion times greater than the entire human race. We lift our finger and a billion cells cooperate with total obedience and precision. When we digest our dinner, billions of complex molecules are rushing about, performing complex chemical tasks.

We have worked long and hard to reach this state, and we have done so by our own efforts. And now the question: what has sustained us in this climb? There can be only one answer. It is sustained by the basic and most fundamental of all the powers, the premonition of a goal implicit in the photon that started it all off. This premonition sustains the quest. It is the thrust, the passion that makes life continually try to excel itself to evolve and, in almost all mankind, has led man to postulate a state of being beyond himself.

Religion does not induce this belief. The belief is in our bones and blood and, when it so chooses, gives sanction to religion. Religion, in fact, is its outward clothing, a method of sharing and articulating the incomprehensible life force.

This brings us to the apparently paradoxical position of seeing religion as the manifestation of physical and emotional, rather than spiritual, causes. But it must be remembered that physical substance and/or energy was itself the product of action, that ultimate totality whose division produces substance and uses it as a vehicle. So the source of the faith in what is beyond oneself is a timeless overview, the same dynamic orientation that has pushed the physical vehicle through its development and that has guided our steps up the ladder of being since the universe first came into existence.

But it might be objected: what of the case when this primal life force fails, as it evidently does when a civilization deteriorates, when a whole people succumb to a newer and more vigorous invading culture, as did Rome and Mexico? The answer lies in the very fact of resignation. Were the life force a true compulsion, a law of nature, it could not resign. It would cling to the corpse as does the force of gravity. Resignation implies volition, and the recognition that the vehicle has become worn out. It implies the willingness to die, to abandon ship and, I insist, take one's chances on finding a better vehicle.

Here the mind boggles. It can go no further, for it says, "If I lose my life, what am I?" What indeed, for with nothing to pinch to assure yourself you are there, you are nothing, *no thing*. But *thing* thou never wert. This thing you've been pinching never was you.

So we should view the evolutionary force in man, and in all life, as the promise of self-transcendence. It is not a compulsive force like gravity, if indeed it is a force at all, but it induces internal transformation. The angiosperms, the most highly evolved of plants, put forth flowers which attract insects. We can call this a mechanism for survival, but the simpler plants survive without flowers. We can suppose, if we wish, that the flowering plants know about insects, or we can insist they do not, but to suppose that the necessity of survival has produced flowers is to deceive our intelligence and betray the spirit of inquiry.

So what should we do? Reexamine the question. Look at the corresponding development in another context. Do the most complex atoms develop radioactivity in order to survive? Is survival the cause of DNA? By looking at the culmination of power in these other areas, we break free of the compulsion to view life as caused by outer circumstances, for the atom is under no necessity to survive. Its development is a pattern of unfoldment. We thus see deeper reasons for evolution. It is adventure, the exploration of possibility, the creation of a game and the play of the game—and not just to win, because to lose lays the basis for a better game. The oysters are still at the game of being oysters. Other creatures lost interest in that occupation and crawled out of their shells, not to survive but to expand their sphere of action.

I trust that this point has been sufficiently proven. The Great Chain of Being is testimony to it. The grid is but another voice ushering the facts of physics and chemistry and the progression of life forms to witness the universal ongoingness of Nature, giving testimony to her ambition to explore all opportunities to create forms that can draw on the nourishment she provides.

But in taking thought about life, in trying to observe it, and its forms, we become isolated from it. We are one of these forms. How may we read our role? For while we too are creatures of nature, actors in the great drama, it is not enough for us simply to exist. We have reached a point where we can begin to write the script.

The arc—reapplied

The arc of process describes the course by which molecules learn to become proteins and DNA; cells to become great trees; amoebae to become a million species of animals—the lion and the antelope it pursues, the hawk and the fish it carries aloft. All creatures have their progression. Ours is to be men, not, like our animal forebears, to perfect a specific function—to run, to swim, or to fly—but to have dominion over nature, to rule it and to care for it, to conceive and attain goals beyond our immediate necessity.

In the construction of a theory of process, we have applied the arc to a variety of things. Let us now apply it to ourselves, especially to the question of the continuity of life, for it is to such unprovable and long-term phenomena that the investment of a theory based on broad principles affords a foundation.

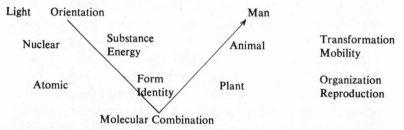

As we have earlier pointed out, the upper two levels are by nature nonfinite, the lower two finite. This follows from ontological necessity, for that which is finite must add boundaries to that which endures, and thus is compound. It requires the extra commitment of a form, and thus depends on the prior existence of that which has extension or duration only. Thus the atom is compounded of nuclear particles— a more primitive existence with duration but no form. When these nuclear particles are separated, the atom ceases to exist. The substance it contains, however, continues because substance (the mass–energy of the physicists) cannot be destroyed.

Applying this schema to man:

  I Monad
  II Soul
  III Mind
   IV Body

We may now define his essential ingredients. At the top level we place that principle in man that is at the core of consciousness. We may refer to it as the *monad* and give it the function of attention, intention, or purpose.

To the second level we may assign the first vehicle of the monad, that which makes possible experience, that by which it feels pain and pleasure, and has the function of memory. We call this *soul*.

To the third level we assign *mind*, self-consciousness, identity. This is the function that makes concept formation possible and, it would appear, requires a body with its nervous system, sense mechanism, and brain.

To the fourth level we assign the physical *body* itself. This includes the physical substance that makes up the body.

This assignment of principles in man implies that what we are calling soul is immortal. It cannot be destroyed. Like the mass energy to which we have correlated it, the soul is without form, it has no parts, and hence is indestructible. This correlation, in fact, is most instructive, for the fluid nature of the soul is what adapts it to taking on the imprint of experience. Like water passed from one vessel to another, it takes the form of the vessel it is in, and supplies the content of experience. The physicist too finds the analogy to water most suitable to describe nuclear particles, whose behavior is often likened to water drops.

The monad too is immortal, but in so describing it we are guilty of an inversion, because the monad is not in time. Here language fails us. I trust, however, that the reader, having come this far, realizes the predicament of describing this essence, which is analogous to the eye that can see everything but cannot see itself; so I will not attempt the impossible. It is necessary, however, to point out that this principle (spirit) is often confused with soul partly because, regardless of words, it is not realized that there are not one, but two, "nonobjective" principles. And let us not be confused by the fact that the words *soul* and *spirit* have reversed roles over the years. There were times when spirit has had the connotation which we would assign to soul (i.e., "earthbound spirit" or "thy father's spirit" used as in Hamlet in the sense of ghost). In any case, what we assign to first level is prior to soul; it is the Hindu *atman*, the *nous* of Aristotle, etc. Soul is the precipitation of spirit into time, the first stage of its "descent" into matter.

Macrobius, who based his view on Pythagorean traditions, thus described the soul's entering into its descent:

> Just as the line is born from the point,
> so the soul from its point, that is monad,
> comes into the dyad, its first lengthening.*

The "lengthening" is duration in time. However, the spirit or monad is not entirely lost in this descent, but remains itself.

The concept is a difficult one. It is, in fact, one of the great mysteries of existence, and has been lost to Western tradition. Aristotle, who inherited from Plato the notion that the soul was an indestructible substance, in later life rejected this view and described the soul as a *formative* principle and not separable from the body.

In reaching that conclusion, Aristotle established the direction which modern thought has taken up to now, putting emphasis on that which can be defined and rejecting the notion of substance. This emphasis has led to logical positivism, which explicitly states that it can deal only with that which can be defined, which is to say, with form.

We have repeatedly stressed the shortcomings of the modern rationalist view, which would eliminate substance (and value) because it cannot be defined. Only substance can supply the chemist with the material for his experiments (as opposed to the information), and the concept of substance is essential to such critical facts as the number of molecules in a gram, or the absolute magnitude of what otherwise is known through ratio.

The concept of energy as used by science conforms precisely to the concept of substance as used by philosophers, and thus constitutes a refutation of Berkeley, who argued that the concept of substance should be dropped. It also reassures those skeptics like Locke who, without trying to reason substance out of existence, remain in doubt about what

---

* Mead, G. R. S. *Thrice Great Hermes* (vol. I, p. 288). London: John M. Watkins, 1906.

cannot be objectively verified—for the answer is that energy, despite its nonobjectivity, does exist.

It then is important to attend to what science has found about the indestructibility of energy. This principle (the conservation of energy) was first recognized in conjunction with the conservation of mass. It was sometime in the 18th century that Carnot, a scientist, vigorously stirred a container of water and measured the amount of energy he thus expended. He then measured the rise in temperature of the agitated water, and found that the energy he had expended had gone into heat. This inaugurated the principle of the conservation of energy. The conservation of mass had previously been established on the basis that whenever a chemical change takes place, the mass of the products is the same as the mass of the initial ingredients. Thus:

Mass of fuel (e.g., hydrocarbon) + oxygen = mass of carbon
dioxide and water

More recently, it was recognized that energy has mass, and that when energy is released from combustion, there is a very small reduction in the mass of the products. This principle had its complete confirmation in the atomic bomb, with the famous equation:

$$\text{Energy} = \text{mass} \times c^2 \ (c = \text{velocity of light})$$

The amount of energy in mass is enormous. The term now used, megaton, measures the energy of atomic explosions in terms of millions of tons of dynamite.

So great is this ratio, however, that for ordinary chemical processes the change in mass is far too small to be measured. The true authority for the principle is not experiment but theory. That which exists cannot vanish into nothing. It was such authority that impressed the primitive mind with the persistence of life after death, and I can see no essential difference between immortality and what science calls conservation.

The importance of this excursion into science will now be clear, for the same reasoning which establishes the substance of physics establishes the substance of the soul. Unfortunately, this may seem to the reader an inversion. Surely, the soul, if it exists at all, is of a spiritual nature not accessible to physical experiment. But this is just what the soul is not. The soul is the counterpoise, the complement of *spirit*. It is that which draws the spirit out, or down.

The intercourse of spirit and substance (or soul) creates the physical universe, for the creative spark can fulfill itself only by engagement with matter (mater = mother). This is the intercourse of the divine parents.

And the concept is not new. It is perhaps the oldest of all. What we are doing in this book is to show that science bears out the truth of this oldest of myths of spirit and matter; for the photon, or initial light pulse, is the nonmaterial partner whose interaction with matter creates activity, whether the activity be the motion of nuclear particles, the energy changes of atoms, the changes in bonds of molecules, or the photosynthesis of plants.

Teleology

But we have also done more. We have shown that all process partakes of this nature and has as its first cause a purposive or goal-seeking thrust which can realize itself only through a marriage with matter.

This justifies our bringing the principle to bear on the nature of man. Man shares with all other creatures and entities an origin in light, but, as a creature which has evolved through all the stages previous to dominion, he takes seniority over those stages, as Genesis puts it:

> Be fruitful, and multiply, and replenish the earth, and subdue it: and have dominion over the fish of the sea, and over the fowl of the air, and over every living thing that moveth upon the earth.

So much for the promise held forth for the dominion kingdom. But as in other kingdoms, the power which makes dominion possible has to evolve.

We recall by what long and involved evolution the mobility of animals was attained. How, in order to gain size, mobility was at first sacrificed (in sponges), how the animal then had to devote its whole attention to creating a stomach, then other organs, then their placement in an organized chain of command, until finally, with the addition of articulated feet, true mobility was at last attained.

Man is doing a different thing, but no less difficult, and the investment he must make in means entails at first a loss of the very freedom that is to be his ultimate reward. We have indicated how we may follow the substages: collective man in substage two, individuation in three, modern man in four, genius in five, etc., but we need to see the whole arc in terms of an organic development, from spirit through the levels of soul, mind, body, and back again through growth, through mastery, to spirit's fulfillment.

This requires realizing that the four levels—spirit, soul, mind, body —are traversed twice: on the way down and on the way up.

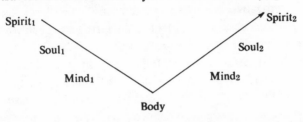

It will have become apparent to the reader a long while back that there is a great difference between the left- and right-hand branches of the arc. $Soul_1$, like the horse of the Chinese metaphor, wanders about eating grass; $mind_1$, like the driver, is intoxicated; and $spirit_1$, like the owner, is asleep.

This is their natural disposition, which is not set right until the owner awakes and by his own efforts, unaided by natural law, does something. *Here is the "turn,"* when the monad, having accomplished the descent, realizes its heritage and the task it has set itself, and turns back toward its home, the celestial world, going through $mind_2$ (genius), $soul_2$ (higher soul) to reach $spirit_2$.

This does not mean to become unworldly. It means rather the intensely practical task of learning the way things work, for that is the reason for descent into matter, to eat of the fruit of the tree of knowledge of good and evil.

The importance of the turn is recognized in almost all the great religions and mythologies. It is often referred to as the birth of the hero. Thus in Egyptian tradition Osiris first undergoes a series of increasingly severe limitations leading ultimately to complete dismemberment. Isis, the sister and syzygy or mate of Osiris, then picks up the pieces and puts them together, with the exception of the phallus, which is lost. Despite this, she conceives and bears the infant Horus, who grows up, defeats Set, and becomes the Sun God. The birth of Horus is the beginning of the resurrection of Osiris—the upward turn of our V-shaped arc.

Now, it is of great interest that the rebirth of Horus is a virgin birth, that the conception is without benefit of the usual male organ. What does this mean? The reader will recall our stressing that the monad must initiate the turn *by its own efforts*, unaided by natural law. The monad at this point becomes *first cause*, which is to say that it draws on the spiritual prerogative that is its heritage from the first level. This self-induced conception leads to the virgin birth (a virgin birth being one that occurs without a prior cause, without a father).

Similarly, Genesis treats of the fall of man (his expulsion from the Garden of Eden). The fallen Adam is redeemed by Christ, who symbolized the spiritual rebirth of man as son of God, crucified in matter, but able to rise again and ascend to heaven, where "he sitteth at the right hand of God."

My purpose in mentioning the Christian teaching and correlating it with Egypt is to point out that both accounts are dealing with the same general thesis that we have used to describe evolution.

The reader will appreciate that it was by the superposition of a number of examples of process that it was possible to arrive at the general theory. Many of these examples I have not used in the present text. But it is important to study all possible sources and not be distracted by irrelevant differences. These ancient myths are extremely rich in content, each throwing light on details of process in general that are obscure in other examples, much as the geology of one continent fills in details of history that are obscure in that of another.

For example, myth was of great help in my understanding of the third stage, identity. In Genesis, Adam is the first stage, the monad. Eve, the "mother of all living," is the second principle, that of desiring or wanting. The third stage begins with disobedience, the eating of the fruit of the tree of knowledge of good and evil. This principle has to do with form and identity. But Genesis links it with disobedience. What is the connection? Clearly, the disobedience is an act of *self-determination*, necessary to both self-consciousness and the sense of identity. This, in turn, adds meaning to "having its own center," the general descriptive phrase applicable to the third stage.

Applying this progression to the self and its evolution, only by self-determination can the self become responsible for its own acts, and only by such responsibility can the self attend to the consequences of action, and thus learn.

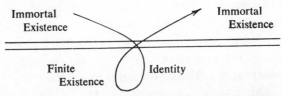

Thus the self, in taking on an identity, becomes responsible to itself and learns how to deal with law in the fourth stage. Having learned the law, it can then act deliberately; it can become cause. This is the turn, and it is followed by the self's "growth."

But the price of identity is finiteness—finiteness in space and time —and this finiteness has a positive value, for were the self not finite, it could not learn. It is because only a self that is finite can suffer from wrong actions and hence learn from experience.

This takes us back to immortality of the soul. We can now reverse the question and ask: if the soul is immortal, why then is the body mortal? It is because only the finite (or mortal) body is appropriate for learning. If the body were indestructible, it could not be injured and no learning could occur. This is the reason for mortality.

The whole system of interrelationships is hereby tied together. The two upper levels are nonfinite, the lower levels finite, and the self "falls" into the mortal realm for very good reasons. It is only by taking roles that the self learns to act, to achieve the competence required in order to have dominion over nature, to become, as Genesis puts it, "as wise as gods, knowing good from evil."

Conclusion

This concludes our testament. We have gone into science and discovered that in addition to its well-explained areas, clearly mapped like the streets of a city, there are portals leading to unknown and unexplained areas, the world of nuclear particles and of light. This is the world of quantum physics in which, contrary to reason, it has been discovered that there is a basic uncertainty, the quantum of action. This is a principle that activates, in contrast to the more accepted scientific principle of law that maintains and regulates.

Thus the creative aspect of the universe revealed by quantum physics confirms the teachings of myth and revealed religion, and hence fulfills our quest for an integration of science and those nonphysical realities thought to be unverified by science.

This is what we have done. We have gone into the inner sanctum of science. We have discovered that it is not billiard balls but *action* which is fundamental. Putting the fundamental entities of science in the precise order of their generation, we have discovered the sequence to be that of the most profound religious teachings.

> In the beginning . . . the earth was without form and void; and darkness was upon the face of the deep. . . . And God said, Let there be light: and there was light.

For the *quantum of action is light*.

But it is not only light; it is directed energy. As it issues from its source, for example the sun, each photon has its precise frequency. Falling into matter, it performs the function that the particular matter requires: if a leaf, the synthesis of starch; if a chromosome, its mutation into a higher form; if on a page of mathematics, the possible illumination of its reader.

How important in this regard is the eye, the quintessence of vertebrate evolution! For the eye is the vehicle most worthy to receive, and most able to benefit from light. Perhaps its nature may help us to comprehend what is perhaps the greatest gap in our theory: the correlation of the quantum of action to the monad.

By the necessary nature of things, *reception* of light must be opposite to light. How can we understand the monad to be at once positive in the sense that action is projective, and also receptive in the sense that consciousness reflects or is aware of existence? There is apparently a dichotomy between spirit and mind, between light and its reception.

The way I have resolved this question in my own mind is to go back to the word used for the perception of truth, to *recognize*. While we may think (cognize) deeply about a problem, we *re-cognize* the solution. Cognition or awareness, which receives the light because it is opposite light, is not the essential activity of the monad. Rather, it is *re*cognition (which is opposite cognition) that is the monad's role. This is the light that dawns when we "see the light." It is a positive creation of light. Thus Creation comes at last to recognize itself.

Appendix I

A brief outline of the theory

1. The universe is a *process* put in motion by purpose.
2. The development of process occurs in *stages*.
3. There are *seven* stages.
4. Each stage develops a new *power*.
5. Powers are *cumulative*; each one retains the powers developed in the previous stages.
6. Powers are evolved sequentially in what are called *kingdoms*.

POWER	KINGDOM
1. Potential	Light
2. Substance	Nuclear Particles
3. Form (identity)	Atoms
4. Combination	Molecules
5. Organization	Plants
6. Mobility	Animals
7. Dominion	(Man)

7. *Arc of process*: The early stages of process take on increasing constraint until constraint becomes maximal, at which point there is a *turn*. The later stages of process see the conquest of the constraints and the development of freedom. Freedom in the first half is random, in the last controlled.

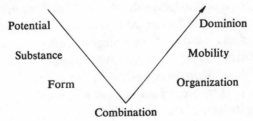

8. *Levels*: The "descent" and "ascent" pass through four levels in a V-shaped arc. Levels have successively zero, one, two, and three degrees of constraint, and three, two, one, and zero degrees of freedom. The stages on the right- and left-hand branches of the arc at the same level have properties in common:

Level I	Purpose	3° of freedom,	0° of constraint
Level II	Substance	2° ″ ″	1° ″ ″
Level III	Form	1° ″ ″	2° ″ ″
Level IV	Combination	0° ″ ″	3° ″ ″

9. *Asymmetry*: The stages on the left and right branches of the arc can be viewed as the inverse of one another:

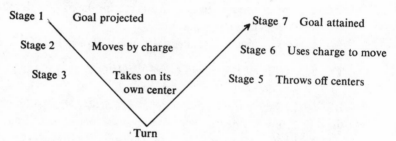

Stage 1 — Goal projected

Stage 2 — Moves by charge

Stage 3 — Takes on its own center

Turn

Stage 7 — Goal attained

Stage 6 — Uses charge to move

Stage 5 — Throws off centers

10. *1-1-2 Pattern*: Each *even* stage begins at the beginning of the previous stage, and each *odd* stage begins at the end of the previous stage.

11. *Self-mapping*: Each stage of process (or kingdom) is itself a process in which the power of the stage develops. The development of this power occurs in stages called "substages," whose description correlates with that of the main stages.

Note: A point that is easily forgotten and which must therefore be reiterated is that atoms, molecules, cells, etc., are not separate things, but expressions of the monad (or the evolving entity) at successive stages of its evolution. At each stage it acquires a new power.

Since the powers are cumulative, a monad cannot achieve cell-ness, say, without having previously mastered molecular combination; it cannot deal with combination (as a molecule) unless it has previously learned individuation (as an atom).

The message of our study is therefore that nothing comes of itself except the initial venturesomeness. This venturesomeness, which started it all off, is always present, pushing process through its stages and acquiring greater and greater competence.

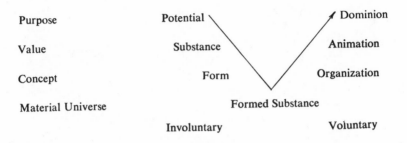

Purpose	Potential	Dominion
Value	Substance	Animation
Concept	Form	Organization
Material Universe		Formed Substance
	Involuntary	Voluntary

The arc: additional properties of the levels

It is helpful to point out certain properties of the arc in addition to the degree of freedom of the several levels, their symmetry, etc.

As we pointed out in Chapter XV, the upper two levels are projective and therefore nonfinite, the lower two levels are objective and finite; viz., the indestructibility (conservation) of mass energy and the immortality of the soul versus the destructibility of forms and the mortality of organisms.

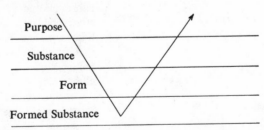

Purpose
Substance
Form
Formed Substance

We can distinguish the first and second level by noting that the first is particular, the second general. The same distinction applies to the third and fourth levels, but in the opposite order, thus: the third level is general (because concepts are general), and the fourth level is particular (because physical objects are particular). The four word pairings can be shown in a diagram.

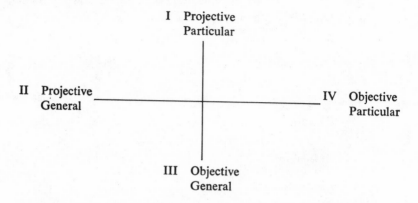

I   Projective
    Particular

II   Projective
     General

IV   Objective
     Particular

III   Objective
      General

Note that the word pairs at opposite ends of each axis are doubly opposite, whereas the word pairs at right angles share one word.

The next property is that the horizontal axis is physical, while the vertical axis, levels I and III, is nonphysical.

For example:

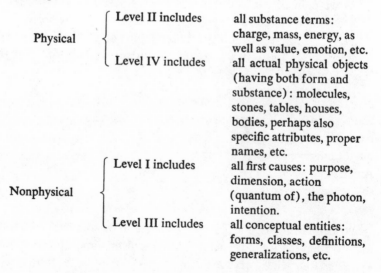

Physical
  Level II includes — all substance terms: charge, mass, energy, as well as value, emotion, etc.
  Level IV includes — all actual physical objects (having both form and substance): molecules, stones, tables, houses, bodies, perhaps also specific attributes, proper names, etc.

Nonphysical
  Level I includes — all first causes: purpose, dimension, action (quantum of), the photon, intention.
  Level III includes — all conceptual entities: forms, classes, definitions, generalizations, etc.

Appendix II
Seven-ness

The intelligent reader is likely to regard with suspicion a theory which bases itself on the number seven.

Why? Is it because any reference to seven conjures up the prescientific? Legend, myth, fairy tales, and even superstitions are encumbered with sevens for no apparent reason and, by giving this number a mystique that the modern mind cannot accept, have given it a bad reputation.

As I stated in the Introduction, my first contact (since nursery days) with the occurrence of this number in ancient cosmologies affected me similarly: it rubbed me the wrong way. However, when I realized that topology, a science that deals with even more profound implications than does geometry, could supply formal reasons for a sevenfold ordering, I was prodded into a rethinking of the concepts embodied in relativity whose theories of curved space–time had provided the foundations for modern speculation on the nature of the universe.

My early curiosity about first principles, coupled with the publicity given, at the time, to the theory of relativity ("only twelve scientists could understand it" was what the press notices gave out), led me to request it as my major. Whether the pupil was ready I shall never know, but the master appeared. He was Oswald Veblen, perhaps the foremost American mathematician of the time, and he willingly took me on as his sole student for the two remaining years of my college career. Each week I would read and copy his manuscript of his version of the mysterious new subject. My other text was Eddington's *Mathematical Theory of Relativity*.* Both placed great emphasis on the "summation

* Eddington, Sir Arthur Stanley. *Mathematical Theory of Relativity*. New York: Macmillan Co., 1923.

convention," which was a mathematical shorthand that permitted one to write countless equations with a few Greek squiggles. Even when I could sometimes get the hang of this painful exercise, I still wanted to know what it all meant, and would constantly harass Veblen for enlightenment. This, Veblen would tell me, was not the business of mathematics. I must learn to manipulate the language; the mathematician should not concern himself with meaning.

Nevertheless, something did seep through, and Veblen in his unguarded moments would occasionally grant me indulgence, enough to keep my curiosity alive.

After leaving Princeton, I continued to ponder on the questions raised by relativity and the adequacy of its answers, and to try to keep abreast of the more significant but less publicized findings of quantum theory; I kept in touch with Veblen, who meanwhile had moved to the Institute for Advanced Study, which he directed. But I could sense that his interest in mathematics was waning; he showed more concern for the lawn. And Bertrand Russell, who once through mutual friends chanced to come to my house in Paoli, and whom I cornered in order to explain my solution to the paradoxes he had bequeathed to posterity, hastened to assure me that he was no longer interested in mathematics, he had transferred his attentions to women. In any case, aside from my occasional visits to Veblen, which I renewed after the helicopter was completed, and courses in logic which Veblen insisted I take before he would introduce me to Goedel, my career has been *ex cathedra* (though I did my homework, Veblen died before he could introduce me to Goedel).

But to return to relativity. Einstein's contribution (or rather, as Veblen used to insist, that of the mathematicians before him, Riemann, Lobachevski, et al.) was to provide a more generalized geometry, a curved space–time, to replace the flat space of Euclid. While this was intriguing, and led to a lot of complicated equations, to my way of thinking, it did not yield anything of great importance. The one significant item, the invariance of the speed of light, was of course highly significant, but this came out of the special theory of relativity, and could even be derived independently. The conclusion of the general theory—that light be bent more than predicted by Newton and that the perihelion of Mercury should advance a few seconds per century—seems small potatoes for such heavy investment. The mountain labored and gave forth a mouse.

More significantly, relativity was a misnomer: the theory was only incidentally a theory of relativity. Its true purpose was the search for invariants, i.e., absolutes: quantities which would be the same for all observers. It is this aspect of relativity that has had the greatest impact and has led to the present so-called cosmological postulate: the thesis that the universe, except for local irregularities, is the same for all observers. Right or wrong, this point of view is central to modern cosmology.

Moreover, relativity, as the search for absolutes, overlooks a very basic and important absolute, namely, *rotation*. (The reader may satisfy himself that rotation is an absolute by getting up from his chair, making a 360° rotation, and sitting down. If rotation were relative, he would be entitled to say that the universe had spun around him, and not that he had turned. But if he maintains this, then all the stars must have moved through space at velocities exceeding the speed of light. Alpha Centauri, the nearest star, would have had to move forty-two light years in the time it took him to turn around, that is, over a billion times the speed of light, and this possibility is barred by the theory of relativity. Therefore rotation is absolute and not relative.)

This question has not been overlooked; P. Bridgeman wrote a rather inconclusive book on the subject. Eddington too, in *The Nature of the Physical World,** considers absolute rotation, but fails to recognize its importance for cosmology, or its ultimate connection with angular momentum, which he does show to be significant as an absolute. Momentum is relative, but angular momentum, based on rotation, is absolute.

In any case, my dissatisfaction with relativity fed my interest in the alternative theory that began to open up after I recognized the theoretical importance of seven-ness, which not only confirmed the possibility of seven stages to process, but cast new light on the old questions. In particular, the conflict between the continuum of relativity and the discreteness of quantum theory—a question with which Einstein had

* Eddington, Sir Arthur Stanley. *The Nature of the Physical World.* New York: Macmillan Co., 1937; Ann Arbor: University of Michigan Press, 1958.

been concerned, but which he did not resolve. Einstein, in fact, could not accept the implication of quantum theory: "God," said he, "does not play dice with the universe."

The torus and individual existence

But I could see a way out. What the quantum theory established are the singularities within the space–time continuum: photons, protons, or electrons; these entities are quasi-independent point sources, islands of uncertainty (or of mass, it matters not). They are unique existences with a definite self-energy. My predisposition to analogies led me to see the problem of their existence as equivalent to the old problem of free will in a universe run by God: how can there be self-determined entities in the continuum? The solution was provided by the torus, which furnishes what is known as a different connectivity. In the case of a plane or the surface of a sphere,

a circle cut around a point *a* will completely separate it from the rest of the surface (1). In the case of the torus, a circle cut around a point need not necessarily separate it (2), for here, despite cut *b*, point *a* remains attached to the rest of the torus. So too, the self, in a toroidal universe, can be both separate and connected with the rest of the universe. And the problem is the same for many selves which would constitute more holes; a hole for each but all connected.

Meanwhile, the evidence for seven stages to process found reinforcement from my study of nature; in the case of atoms, the periodic table showing seven rows; in the case of molecules, seven orders of combination, and so on. The theoretical evidence also grew. To my great excitement, I discovered that the formula for the volume of the

Einstein–Eddington universe, the so-called hypersphere, was $2\pi^2R^3$, the same as the volume of the torus with an infinitely small hole!

Eddington discusses the hypersphere in his *Fundamental Theory*, in which he answers the question of the unification of quantum theory and relativity once and for all. (See Appendix III.)

Eddington explains that the new way of taking account of curvature is *phase*, and that phase space will constitute a fifth dimension at right angles to space–time.

Phase is, of course, angular relationship and has a range of 360° or 2π,* because angle is measured over this range. It can be direction, or it can be what I called *timing* in Chapter V, the lead or lag between an input and a reaction.

360°-45°   45°

Maximum angle $= 360° = 2\pi$

Since *choice* is expressed either as direction or as timing, *it can be equated to the phase dimension.* That is to say, if we want to assign a dimension to choice, the appropriate dimension would be angle. (The question of whether you vote with [in phase] or against [out of phase] is one of angle.)

To carry the argument a step further, our uncertainty of what someone else would choose would also be expressed as an angle, and the maximum uncertainty as the angle 2π. Quantum physics tells us that our uncertainty in a given situation has the value h, a unit of action. This unit, since it can be divided by 2π to obtain h bar, must contain 2π, and hence includes uncertainty of direction (as well as of action).

This can be made clearer if we realize that, besides the limitations described by Heisenberg, observation is confined to its own cycle of action. For example, we hear a sound. If the frequency of the sound is reduced to much below 16 cycles per second, we no longer hear it as a sound; it becomes a series of separate beats or is heard as a rattle.

* Strictly 2π radians where a radian is a radius laid out on the circumference of a circle.

Contrariwise, if we were counting beats and the frequency passed above 16 cps, we would no longer distinguish separate beats; from that point on, we would hear a steady sound, and be unable to distinguish fractions of a cycle. Our measurement would be $n \pm \frac{1}{2}$, which in terms of cycle would be $n \pm \pi$, i.e., an *uncertainty* of 2π (as we've said, you might count the turns of the corkscrew, but you wouldn't know where it ended up).

This uncertainty holds for all measurement, it being understood that the uncertainty is that of the observer while for the object it is freedom.

We now have the pieces of our puzzle and can put them together. We have two contending theories to reconcile, relativity and quantum theory. Relativity says that the universe or space–time is curved and continuous. Quantum theory says it is discontinuous. Relativity says that curved space–time might be spherical or it might be saddle-shaped.

a. Spherical shape

b. Saddle shape

In one case, parallel lines converge; in the other, they diverge. Of the two, the latter is the more interesting. We say if space–time is to be continuous and saddle-shaped (say), then it cannot stop as is shown in the sketch because if it did, it cannot have a boundary without requiring a higher space in which it is imbedded. Therefore it must continue. What will we get if we fill in the part that is not shown?

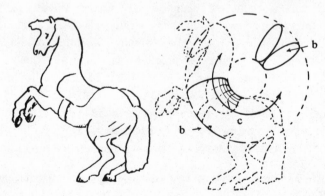

Thinking of the part of the saddle that goes around the girth of the horse and filling this in, we will obtain b. Thinking of the front and back continued in the direction of the arrows, we will obtain c. Thus the whole figure will produce a torus or doughnut of which the original saddle space is only a part.

Quantum physics and the control phase

What about quantum physics? As Eddington pointed out, the curvature of space–time can be replaced by the phase dimension whose measure is 2π, and this 2π is the uncertainty inherent in quantum theory. In other words, Eddington recognizes that the curvature of relativity is the same thing as the uncertainty of quantum theory! But how about scale? Certainly the microscopic uncertainty of an individual proton is not the same as the vast curvature of space–time, which wouldn't "return" or complete its cycle for billions of years. True, but both have the same topology! Whether we deal with the particle or the universe of particles, the topology is toroidal;* their difference is in their time scale.

The torus

Returning to the saddle space of relativity, it can now be appreciated that the extension of the saddle which carried it around the girth of the horse, and again in a vertical plane to complete a torus, constitutes two circularities, *two* uses of π, and these two π's bring it about that:

the volume of the torus is $2\pi^2 R^3$
and the surface is $4\pi^2 R^2$
We therefore say that the hypersphere is a torus!

* This, incidentally, is the conclusion reached by James Archibald Wheeler on rather different grounds in *Geometrodynamics*, New York: Academic Press, 1974.

It is curious that this hasn't been recognized (I asked Dr. Wigner at Princeton about this, and he thought it had not been noticed). In any case, the torus and the hypersphere have the same formula, and the requirement that space–time be both saddle space and continuous can be met only by the toroidal figure.

The next step is more drastic. We want to know what the extra π means. We know that it has to do with the cycle of action. That is, we know that every entity, from atom to galaxy, has a cycle of action and this cycle is represented by 2π. But is this all? If we recognize that anything physical expresses itself spatially as a sphere—in the sense that a sphere is the "shape" of a particle or it is the radius of action of a fly or a man—and that the sphere has a volume $(4/3)\pi R^3$, we cannot simply multiply this spatial existence by 2π. The product would be $(8/3)\pi^2 R^3$, which differs conspicuously from $2\pi^2 R^3$. Something is wrong; we are off by a factor of 3/4. (Eddington discusses this factor of 3/4, but since I cannot follow his reasoning I will give my own and refer the reader to his *Fundamental Theory*.) (See Appendix III.)

This factor of 3/4 is the most interesting of all. For it so happens that it is the 3/4 point of the cycle of action that has significance. Let us consider the different phases of the learning cycle, which is one example of the cycle of action.

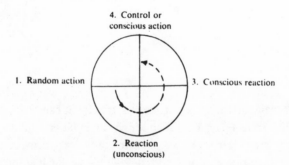

4. Control or conscious action

1. Random action

3. Conscious reaction

2. Reaction (unconscious)

The start of the learning cycle is blind action: the infant reaches out at (1) and touches the hot stove, and at (2) reacts. Then at (3) he considers what has occurred; he becomes conscious that hot stoves hurt (if you are a behaviorist you can say he associates the pain with the stove), then at (4) he *avoids* the hot stove. This is the 3/4 point, conscious action and merges back into (1) as avoidance becomes instinctive.

However, it is the 3/4 point that is important. For this is the point of conscious choice. It makes possible the "turn," first discussed in Chapter V and, as was there shown, its dimensions are $L \times T$ or L/T^3. (This is the third derivative of position, i.e., change of acceleration or control, the same control we use in driving a car. What we control is force, for force is required to accelerate the car.)

$$L \times T \times \text{force, or } LT \times ML/T^2 = ML^2/T = \text{Action}$$

but also:

$$3/4 \times 2\pi \times (4/3)\pi R^3 = 2\pi^2 R^3$$

In other words, the extra 2π makes control possible. It is the entry of consciousness into the universe. Its existence is borne out by the formula for the hypersphere, by the extra uncertainty in the quantum of action (which does not become control until it reaches the turn), and by Eddington's profound recognition that the curvature of relativity is equivalent to the uncertainty of quantum theory. We add that both are the capacity of consciousness to act upon the universe, or to control determinism.

The seven postulates of projective geometry

These were the theoretical considerations that poured in upon me after I recognized that the universe and the creatures which inhabit it are toroidal. But while I now had plenty of evidence to show that the universe is toroidal, I did not have proof that the seven colors required to map the torus could be correlated to the seven stages of process.* It was the quest for this that made me sit up and take notice when, after Veblen's death, I read in an article entitled "A Mathematical Science"** that Veblen, in cooperation with John W. Young, had set forth the postulates (which they call assumptions) on which projective geometry

* See p. xxi and *The polyvertons*, later in this appendix.
** Newman, James R. *The World of Mathematics*. New York: Simon & Schuster, 1956–1960.

is based. Projective geometry is the science that deals with the properties of figures as they are seen from different points of view by an observer; it is more general than ordinary or metric geometry, in which shapes do not change.

Ordinary geometry is based on four postulates, so imagine my excitement to discover that Veblen and Young had found that for projective geometry *seven* postulates were required:

 I. If A and B are distinct elements of S, there is at least one class containing them.

 II. If A and B are distinct elements of S, there is no more than one class containing them.

 III. Any two classes have at least one element of S in common.

 IV. There exists at least one class in S.

 V. Every class contains at least three elements.

 VI. All elements in S do not belong in the same class.

 VII. No class contains more than three elements.

I immediately set to work to arrange these seven postulates in an arc. It was evident that they paired off in two sets, one saying "at least" and the other "no more than." Further, two dealt with one-ness (IV and VI), two dealt with two-ness (I and II), and two with three-ness (V and VII), leaving number III to do with combination for the bottom of the V (arc).

	At least	*No more than*
Post. IV	One class in S	VI All elements not in one class
	I Two elements in a class	II No more than one class for two elements
	V Three elements in a class	VII No more than three elements in a class
	III Two classes have a point in common	

Placed in this manner, we can see that the postulates obey the rules of symmetry developed for the arc: one-ness, two-ness, and three-ness on the successive levels, with combination on the fourth, and the right-hand side inverting or reversing the left-hand side. There is even a resemblance in meaning, for the right-hand side, "no more than," exhibits a self-restraint that contrasts with the "indulgence" of the left-hand side. This demonstrates a correlation between the postulates for projective geometry and the stages of process. We now need to show a correlation between the mapping problem and the postulates.

The polyvertons

To do so, we can begin by making a closer scrutiny of mapping. Mapping, by which we here mean coloring areas (countries) of a map in such a way that no two adjacent countries have the same color, seems a rather trivial subject, not entitled to the deep significance which attaches to topology which, as is well known, involves a more profound type of relationship than can be expressed in geometry.

But the coloring problem is not trivial. To realize this, consider the problem of the number of points that may be joined each to each by lines which do not cross.

As the reader may verify, no more than four points on a plane surface may be joined.

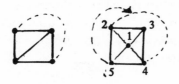

Five points cannot. We cannot reach 5 from 3 without crossing a line.

This may seem no less trivial than mapping, for it is, in fact, the same problem in another guise, the line which joins two points being equivalent to a boundary between two areas and hence to a color distinction. Since coloring can be thus generalized, it can hardly be trivial.

Let us now impose the requirement that the lines must be the same length. We start with four points in a square as before:

But the line 1–3, being a diagonal, cannot be the same length as the sides. In order for it to be the same, we must distort the figure, making in effect two equilateral triangles. But now we have increased the distance between 2 and 4. To join 2 and 4, we must hinge the triangles out of the plane of the paper and obtain a solid figure.

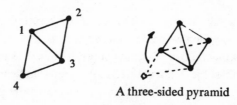

A three-sided pyramid

This is a tetrahedron. It has four vertices, six edges, and four faces. It is the figure which results from joining four points, each to each, with lines of the same length, and requires three dimensions.

In similar fashion, the equilateral triangle is the figure which results from joining three points with lines of the same length and requires two dimensions.

The unit line joins two points and requires one dimension, and the point, which stands alone, requires no dimensions.

We now have four figures correlating with four dimensionalities. The association with dimension makes it possible to recognize that these abstract entities have a resemblance to the powers of the first four kingdoms because they have the same number of dimensions as the kingdoms have constraints.

We can also recognize a correlation of the point to the potential power; of the line to binding, which ties things together; of the plane to form, which requires two dimensions to portray; of the solid to material objects, which combine form and substance.

But this is preliminary. We want to extend the method to deal with the higher dimensions and so obtain information about the higher

Figure	Dimension	Dimension	Constraint	Power
•	0	Point	0	Potential
•——•	1	Line	1	Binding
△	2	Plane	2	Form
◇	3	Solid	3	Material objects

kingdoms; the method of joining a number of points each to each gives us something to hold on to in this difficult area.

Beyond three dimensions

We now inquire: what happens when we join five points? Obviously, we cannot do so in three dimensions without stretching one line.

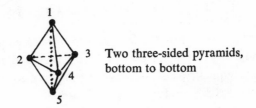

Two three-sided pyramids, bottom to bottom

We can join every point to every other point up to the last pair, shown in the figure as the vertical and internal diagonal 1–5. This line must be stretched.

Note, of course, that this is not one of the regular polyhedrons. It has five vertices, and *apparently* nine edges and six faces. But we are forgetting the interior diagonal 1–5, which adds another edge, making *ten* edges. This diagonal also contributes to three internal faces, which along with the face 234, makes ten. These numbers are familiar. They are the coefficients of an equation of the fifth degree, or $(x + 1)^5$:

$$x^5 + 5x^4 + 10x^3 + 10x^2 + 5x + 1$$

We can see that the first 5 enumerates the vertices, the first 10 the edges, the second 10 the faces. What is the final 5? It is the number of tetrahedrons which can be obtained by selecting any four of the five vertices:

$$
\begin{array}{ccccc}
1 & 2 & 3 & 4 & - \\
- & 2 & 3 & 4 & 5 \\
1 & - & 3 & 4 & 5 \\
1 & 2 & 3 & - & 5 \\
1 & 2 & - & 4 & 5
\end{array}
$$

This alerts us to the fact that the point, line, triangle, and tetrahedron are also described by the coefficients of equations of lower degrees: this build-up of numbers is known as Pascal's triangle.

1 1	$= (x+1)^0 = 1\ 1$	1 point
1 2 1	$= (x+1)^2 = 1\ 2\ 1$	2 points, 1 edge
1 3 3 1	$= (x+1)^3 = 1\ 3\ 3\ 1$	3 points, 3 edges, 1 face
1 4 6 4 1	$= (x+1)^4 = 1\ 4\ 6\ 4\ 1$	4 points, 6 edges, 4 faces

The correlation of coefficients to vertices, edges, etc., emphasizes the more abstract significance of these figures, and in addition, provides us with a quick way of knowing the number of edges, faces, etc., which becomes difficult for the higher figures.

What can we call these figures? Polyhedron, which means "many-sided," is not the right word, since we would like to include the triangle, the line, and the point, which have no sides. But all the figures have vertices, so we may call them *polyvertons*. The figure having five vertices becomes, then, the *pentaverton*.

We noted that it could not be constructed in three dimensions without stretching one line. But since a stretched diagonal is an internal stress which implies storage of energy, we may correlate the pentaverton with the fifth kingdom, plants, for this is what the plant is able to do. So we now say a *seed* is a *four-dimensional object* (or five-dimensional if we have the zeroth dimension).

This promises to remove higher dimensions from their usual employment, which is impossible to visualize. We now have given concrete confirmation, for we can *see* a seed. It externally resembles any other solid object, say a pebble, but it contains an internal organization having the power to store and expend energy.

Joining six points gives us the hexaverton.

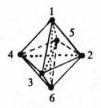

It has some resemblances to an octahedron, since it has eight exposed sides, each one an equilateral triangle. However, when we join all points we get fifteen edges. How many faces? The formula gives:

1	6	15	20	15	6	1

So there must be twenty faces and fifteen tetrahedrons (see if you can find them!).

Here we must have *three* internal diagonals,* each one stretched. This hexaverton correlates to the animal kingdom. The three stretched diagonals suggest the animal's power to change its shape by stretching in different directions, which is the basis for animal motion.

A further property of this model is interesting. Its description takes us back to mapping or its equivalent, joining points on a surface.

How may we join six points with lines that do not touch? This cannot be done on a plane surface, which is inadequate for more than four points. However, it can be done on a Moebius strip.

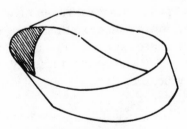

Moebius strip

* 1–6, 5–4, 3–2.

The tracing of the figure is facilitated by the convention of showing a Moebius strip as a square and remembering that the opposite edges join in an inverted fashion:

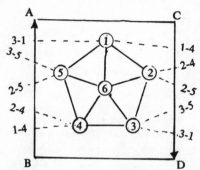

Note that the edge AB joins DC so that B touches C and A touches D

The right edge of the square is pictured as going around the back, twisting one half a turn, and joining the left. This twist makes it possible to join each of the five points to every other point. (The sixth point at the center is joined to all five which surround it.) Note that each point has five lines leaving it.

This seems quite different from mapping, but it is really the same problem. A map on the Moebius strip can be made which requires six colors.*

The heptaverton

Connecting seven points each to each requires twenty-one lines or edges. Starting with (1), we draw lines to the remaining six. We then join (2) to five more—it is already joined to (1)—and (3) to four more, and so on. So we have $6 + 5 + 4 + 3 + 2 + 1 = 21$. We can confirm this from the coefficients, which are:

$$1, \quad 7, \quad 21, \quad 35, \quad 35, \quad 21, \quad 7, \quad 1$$

* Contrasted to the seven required on a torus, as we pointed out earlier in this appendix.

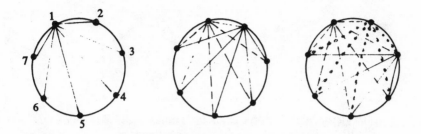

This figure can be thought of as adding a point at the center of the octahedron, and this additional point creates a set of six "compressed" diagonals in addition to the fifteen edges of the hexaverton, making twenty-one.

That this figure is the equivalent of the seven-color map is evident from the fact that seven points may be connected each to each on the surface of a torus with no intersections.

The demonstration is similar to that for the hexaverton on the Moebius strip, except that we now represent the torus as a plane whose opposite edges are imagined to curve around and join, top to bottom and right to left. Here there is no twist, but both pairs of edges join.

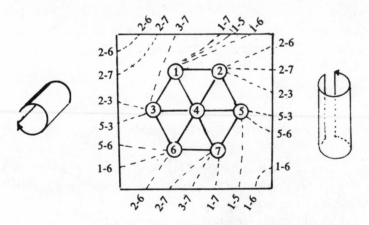

What is most interesting is that with the heptaverton we have the maximum possible number of triangles around each vertex. The tetraverton has three, the pentaverton four, the hexaverton five:

Since in the heptaverton each point connects with six others, the six resulting equilateral triangles fill up the 360° of space available in a plane. *This suggests we can go no further.*

A quite interesting paradox is suggested by the fact that space would be filled by the heptaverton. Suppose we were to make a heptaverton model of the universe. We would find that the interface between the model and the universe extended infinitely in all directions. And we would not know whether we were inside or outside the model!

While the vertons beyond the tetraverton (four points) are not possible in three dimensions of ordinary space and require the addition of imaginary dimensions, it would appear that joining more than seven points is impossible even with imaginary space.

This is given some credence by the progression we have encountered, first one, two, and three real dimensions, then one, two, and three imaginary dimensions.

Counting the zero dimensionality of the point, we have seven dimensions, and if these are all that are possible, we confirm the sevenfold nature of process.

Seven postulates compared to seven points

It is interesting that the problem of joining seven points each to each is in some respects equivalent to the postulates of projective geometry mentioned earlier in this appendix.

Veblen and Young show that the postulates imply that there are precisely seven elements or points in S (the totality), and that there are seven classes each of which contains three elements. They illustrate this by posing the problem of creating seven three-man committees from seven men, the committees to be chosen in such a way that no two persons are together in more than one committee (postulate II).

If the seven men are a b c d e f g, we may select the committees by taking a, b, and d for the first committee, b, c, and e for the second, and so on.

```
a b c d e f g
a b – d
  b c – e
    c d – f
      d e – g
        e f – a
          f g – b
            g a – c
```

This can be shown for the seven points of the heptaverton thus:

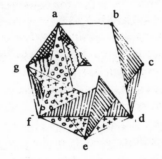

There are seven triangles, each of which contains three points, such that no two points are in more than one triangle (or no two triangles share an edge). The seven triangles use up all twenty-one edges of the heptaverton.

The "classes" of the postulates are thus faces, each one determined by three points.

We can also show that there are seven tetrahedrons with no face in common.

Veblen and Young prove that if two universes S_1 and S_2 satisfy the seven postulates (in that their elements and classes correspond), then S_1 and S_2 are abstractly equivalent, and constitute the same universe. They entitle the set of postulates which lead to such isomorphic examples *categorical*.

We have shown that the heptaverton is:

1. Isomorphic with the torus, in that the connections of seven points with lines which do not cross is equivalent to mapping seven countries. (This is also stated by Hilbert.) *
2. Isomorphic with the seven postulates, in that both their elements (points) and classes (triangles) can be placed in correspondence.

Therefore, since both the torus and the seven postulates are isomorphic to the heptaverton, they are isomorphic to each other. We could therefore conclude that the seven postulates require a toroidal universe.

However, this is not our goal. We want to show that the seven powers, as charted on our grid, are isomorphic to the torus, and this requires that the powers be isomorphic to the postulates.

It should be sufficient to show that the powers can be assigned in one and only one way to the postulates. Since the powers are cumulative, they have a necessary order which of course is that of the kingdoms of nature. We have shown that when the postulates are placed in the order shown on page 268, they correspond to the powers, and preserve the same symmetries. This ordering placed the postulates "at least" on the left (what comes first) and "no more than" on the right (what follows). This is necessary since we cannot limit the membership until it has been filled. The ordering also places postulate IV (there exists at least one class) first. This is the only postulate that could come first, because it establishes existence. Postulate I must follow because it distinguishes two things, and V must come third because it deals in three. The ordering of the last three is determined by the symmetry.

We can thus conclude that the powers have the property of the postulates, and are categorical. Moreover, it follows that any universe, by which we imply any process, must imply the same powers.

Here we must recognize that the description of the powers is subject to considerable flexibility. In a sense, the powers are undefined terms. In another sense, they are not. Their proper definition requires only that:

1. They be independent and categorically distinct.
2. They be cumulative (which implies a definite order).

* Hilbert, D., and Cohn–Vossen, S. *Geometry and the Imagination*. New York: Chelsea Publishing Co., 1956.

These requirements can also be met by distinctions based on dimensionality (i.e., one-ness, two-ness, etc.), which, it will be appreciated, imply degrees of freedom and hence levels (as in the arc).

We are thus equipped with a paradigm that can be enforced as a categorical principle.

The reader is likely to object and say that this violates the principles of valid inquiry, which should be based on the inductive method.

But this misses the whole point of our method. A true scientific inquiry must be based on some paradigm, and if this is not recognized the paradigm remains unconscious. Such is the fourfold paradigm which structures "objective" science and leads to determinism. It is false because it overlooks first cause and makes rationality the final arbiter. Its emphasis on objectivity omits the projective aspect and leads to a cul-de-sac. It fails to discover the reflexive quality of the universe. The sevenfold paradigm corrects these errors and opens the closed system that the fourfold would dictate. It frees us from a misinterpretation of the constraint of law.

Approaching a subject with the sevenfold paradigm can be compared to approaching a machine with a knowledge that, in addition to its structure, it has a purpose and can be turned on, whereas the objective paradigm would permit only a study of its structure.

Importance of the committees

The committees, or subsets of three points, are important because they meet the requirements of the postulates of Veblen and Young. We have shown that the postulates can have a correspondence to the seven kingdoms.

We are entitled to extend this correspondence and ask, "If the postulates correspond to the kingdoms, what do the committees correspond to?"

Clearly, the committees must be interactions between kingdoms—or perhaps better—between the powers of kingdoms.

But my effort to discover such correspondence was not successful until I discovered that the committees could be constructed in a different way from the one already shown.

For example, instead of the connection in Fig. 1, we may connect the points as in Fig. 2.

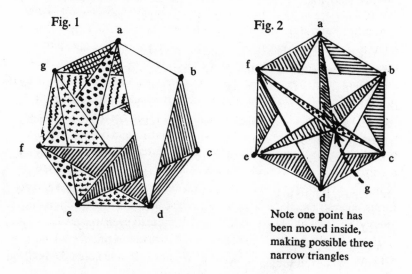

Fig. 1

Fig. 2

Note one point has been moved inside, making possible three narrow triangles

But this was still not fruitful. What did prove fruitful was to recognize that the formation of committees is *cumulative*. As the number of elements increases, the number of committees does too, but more rapidly, and drops to zero after seven. (After seven it becomes impossible to comply with the postulate that any two committees have a member in common.)

Number of elements	1	2	3	4	5	6	7	8
Number of committees	0	0	1	1	2	4	7	0

Taking six points, for example, two more committees become possible when the sixth point is added.

At this time I was working on the material in the chapters on evolution and had drawn the three diagrams shown at the end of Chapter XII, to represent three types of evolution. I then realized the three kinds of evolution were committees—*re* interrelationships between the powers of three kingdoms.

Since there are seven committees, and only three types of evolution, I had four triangles (or committees) unaccounted for. But the problem was simplified by the fact that having filled in the three evolutions, there was only one way to draw the remaining committees, keeping in mind that the first committee consisted of light, particles, and atoms—since this committee would come into existence first. The next committee, in addition to the two evolutions, is not possible until animals, and this committee interrelates cellular organization, animals, and light. I suspect it has to do with the introduction of choice into animals—and choice must have its origin in the free orientation which characterizes light. Surprising to me was the fact that I could not get a committee made up of plants, light, and molecules, which I thought would fit the storage of light by chlorophyll and hence be necessary to the plant kingdom.

The two remaining committees which come into existence with the seventh kingdom are (1) dominion, form, and animation, and (2) dominion, energy, and organization. Both seem to describe distinctly human activities, but I find them hard to distinguish. The animation of forms suggests mechanical gadgets—but so does the organization of energy. So I will have to consider there is more to be discovered.

However, the recognition of the cumulative factor uncovered more evidence for the uniqueness of seven. We have already noted the build-up of committees, which we will now call triads.

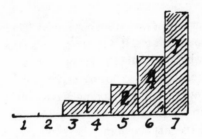

Horizontal scale:
number of elements
Vertical scale: number of triads possible for n elements

In addition to triads, it is possible to have tetrads (four-element subsets). These must have no more than two elements (an edge) in common. As with triads, this requirement cannot be met with more than seven elements.

In the case of the tetrads, there is an even more rapid build-up.

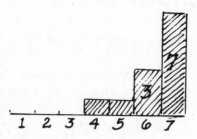

Adding up subsets of n elements we have:

	Number of elements							
	1	*2*	*3*	*4*	*5*	*6*	*7*	*8*
Triads	0	0	1	·1	2	4	7	0
Tetrads	0	0	0	1	1	3	7	0
Grand total of subsets	0	0	1	2	3	7	14	0

In short, then, *the number seven is unique in that it contains the maximum number of possible subsets*—more than twice as many as any other number. It is also the largest number for which linked subsets are possible.

The uniqueness of seven is thus confirmed. There is no other number which has as many subsets. It is interesting also that while we do not at present know what these subsets "mean" or how they apply to practical questions, we believe them to be *interactions between powers*—and hence important to the understanding of cosmology.

Since the importance of subsets does not begin until the fifth kingdom or stage (this is the first kingdom to have two three-element subsets), we might suspect that one of these subsets had to do with evolution because this is where evolution in the usual sense of "survival of the fittest" begins. However, this conclusion was reached independently in Chapter XII.

But why have subsets been overlooked in the exact sciences? I believe the question answers itself. The exact sciences cover only the first four kingdoms in which the notion of subsets is not appropriate (since there is at most only one subset of each kind in a kingdom).

We have thus confirmed, from considerations of the most abstract sort, the uniqueness of seven.

Appendix III
The phase dimension*

The usual equations of wave mechanics postulate flat space. I do not think that there is anything to be gained by trying to extend wave mechanics to curved space. Curvature and wave functions are alternative ways of representing distributions of energy and momentum; and it is probably bad policy to mix them.

We have introduced the curved space of molar relativity theory as a mode of representation of the extraordinary fluctuation, and have obtained the fundamental relation (3 · 8) between the microscopic constant σ and the cosmological constants R_0, N. Having got what we want out of it, space curvature no longer interests us; and we return to flat space to pursue the specialized development of microscopic theory. That does not mean that henceforth we neglect curvature; we merely refrain from using the dodge that introduces it. The scale uncertainty, instead of being disguised as curvature, will be taken into account openly; so that there is no loss of rigour.

Accordingly the scale is now treated as an additional variate whose probability distribution is specified along with that of the ordinary momenta and coordinates. The variates of a probability distribution occur in conjugate pairs, and the variate conjugate to the scale will be called the *phase*. Since we have to provide for cases in which the scale reduces to an eigenvalue, the scale is classed as a momentum and the phase as a coordinate. The phase coordinate is represented as a fifth dimension normal to space–time (which is now flat), so that the scale

* From Eddington, Arthur Stanley. *Fundamental Theory* (pp. 46–47). London: Cambridge University Press, 1946.

and phase are invariant for the rotations and Lorentz transformations of special relativity theory.[a]

The scale uncertainty is primarily a fluctuation of the extraneous standard. But fluctuations of the standard are reflected in the measured characteristics of the system. The scale momentum is the measure of a characteristic which we may call the *scale-indicator*; it is itself unvarying, but its measure shows these reflected fluctuations. In the ordinary momenta the reflected fluctuations of the standard and the fluctuation of the characteristics themselves are inextricably combined; so that we have to introduce one unvarying characteristic to exhibit the scale fluctuation by itself.

We have employed a comparison particle to embody the extraneous standard, and have 'perfected' the object-system by including the comparison particle within it. The introduction of the scale and phase dimension is an equivalent way of perfecting the object-system; and the scale-indicator is the form taken by the comparison particle when it is brought into the object-system. It is a common practice to use a 6-dimensional space to represent a system of two particles. Here one of the particles is a comparison particle, and we need only to extend the object-space by one dimension. Moreover, since the object-system has always to be considered in conjunction with an extraneous standard, the extra dimension is a permanent feature of its representation.

To represent the extraordinary fluctuation or cosmical curvature the scale momentum must be given a Gaussian probability distribution with standard deviation σ. For most purposes this would be a pedantic refinement; and the scale may be regarded as a stabilized characteristic. But now that each particle or small system has its own scale variate, a new field of phenomena is opened to theoretical investigation, which is suppressed in the molar treatment of scale as an averaged characteristic. As remarked in § 23 the comparison particle to be introduced into a microscopic object-system is an individual; and the fluctuation of its energy is of order 1, in contrast to the mean comparison particle whose fluctuations are of order 10^{-39}. We have therefore to distinguish two steps: the substitution of an explicit (5-dimensional) for a concealed

[a] This is not the same as the fifth dimension introduced by curvature. In § 6 the scale was represented by a distance $O'P'$ in the u direction; but distances normal to space–time now represent phase, the scale being a momentum.

(curvature) representation of the mean scale, and in the explicit representation the substitution of individual scales for the mean scale. Since the mean scale is practically a stabilized scale, the second step is described as *the de-stabilization of scale*.

For some purposes it is convenient to take an angular momentum as extraneous standard, so that the scale momentum is an angular momentum and the corresponding phase coordinate is an angle.[a] This facilitates the stabilization of scale—or rather it facilitates the de-stabilization of the fixed scale commonly assumed. The feature of an angular coordinate is that 'infinite uncertainty' corresponds to uniform probability distribution between 0 and 2π. Thus, if J is an angular momentum and θ the corresponding angle, as the uncertainty of J diminishes θ tends to a uniform distribution over the range 2π; and we pass without discontinuity from an almost exact (observed) value to an exact (stabilized) value of J. Conversely, results which assume an exact scale are extended to a slightly fluctuating scale by spreading the distribution uniformly over a thickness 2π in an extra phase dimension. We call 2π the *widening factor*. From the widened distribution we can pass continuously to distributions in which the variation of scale becomes of serious importance.

The widening factor must be taken into account when we compare spherical space (with stabilized scale) and flat space (with fluctuating scale). When the scale is stabilized we have a spherical space whose total volume is $V = 2\pi^2 R^3{}_0$. Preparatory to de-stabilisation this is to be re-ordered as a volume $V_3 = \pi R^3{}_0$ of three-dimensional space having a thickness 2π in an extra phase dimension. Comparing it with a flat sphere of radius R_0 and volume $V_4 = 4/3\pi R^3{}_0$, we have

$$V_3 = 3/4 V_4.$$

Since V^{-1} in natural units is a mass m, this is a relation of the form

$$m_3 = 4/3 m_4,$$

and is an example of the law (16·5) connecting masses of different multiplicity. In V_3 the scale is still exact and the phase necessarily has uniform distribution over the thickness 2π; the representation does not

---

[a] The complete momentum vector contains both linear and angular momentum, so that there is no incongruity in this choice.

give any extra freedom. In V_4 the scale is de-stabilized and the constraint is relaxed, so that the number of degrees of freedom is raised from $k = 3$ to $k = 4$. Conversely, starting with the volume $4/3\pi R^3{}_0$ of flat space, we multiply it by a thickness 2π in the phase dimension, then multiply by $3/4$ to stabilize the scale since the stabilization reduces the number of degrees of freedom from 4 to 3, and so obtain the volume $2\pi^2 R^3{}_0$ of scale-stabilized space which is three-dimensional but curved.

Index

About the Author

Arthur Middleton Young, the son of Eliza Coxe and the Philadelphia landscape painter Charles Morris Young, was born in Paris in 1905. Educated at Princeton, he decided as a young man to devote himself to philosophy. In 1929, however, he embarked on the development of a helicopter, a concept that had long fascinated inventors. For twelve years, he worked alone, using small models to test his ideas. Then, in 1941, the Bell Aircraft Company agreed to build full-scale experimental helicopters to his design, and operations were transferred to the Bell plant at Buffalo, New York. Eventually, on 8 March 1946, Young's machine, the Bell Model 47, was certified by the Civil Aeronautics Board, receiving the first commercial helicopter license ever issued.

Deeply disturbed by the development and use of nuclear weapons, Young came to an awareness that a new paradigm of reality capable of restoring a sense of meaning to human endeavors was an urgent necessity. To this end, he established the Foundation for the Study of Consciousness in 1952 in Philadelphia. Twenty years later, this was followed by the Institute for the Study of Consciousness, located in Berkeley, California. In 1972 Arthur Young and Dr. Charles Muses brought out a collection of essays entitled *Consciousness and Reality*. Youngs's key books *The Reflexive Universe* and *The Geometry of Meaning* appeared in 1976, and extracts from his diaries for the years 1945–48 were published as *The Bell Notes: A Journey from Physics to Metaphysics* in 1979. A further collection of Young's writings were issued in 1980 under the title *Which Way Out? And Other Essays*.

In 1984 Arthur Young's successful helicopter the Bell-47D1, "an object whose delicate beauty is inseparable from its efficiency," was placed on exhibit as part of the permanent collection of the Museum of Modern Art in New York.